No Canary in the Quanta:

Who Gets to Decide
If the Large Hadron Collider
Is Worth Gambling Our Planet?

By Harry V. Lehmann

No Canary in the Quanta: Who Gets to Decide if the Large Hadron Collider is Worth Gambling Our Planet?

By Harry V. Lehmann

Published by Green Swan

Novato, California

Table of Contents

This book is dedicated to Jean Marie Bowler

Acknowledgements

Those of us who read a great deal have noticed a recent tendency amongst some authors to give chatty thanks not only to those who helped with the book, but to those who helped with life, prior to a book. Coffee shops and bookstores sometimes make the cut. Since the people who have helped me in life have been many, and the people who have contributed to this effort have been far fewer, I will mention only the second category.

I thank my spouse Jean Bowler for her support of this project, and in particular for her review of the manuscript, a Hoya in the mix. I thank the indispensable Vicki Clarke, who, with her Masters in Library Science and prior long background as a programmer was the single most contributive member of the team. I thank the following additional people who read the manuscript during its earlier versions, and made useful suggestions: Tom La Flamme, Paige Lehmann, and my friend the unnamed yet famous scientist, whose comments helped with focus. I thank my son Jesse Lehmann for his adamant position regarding the moral issues presented by the multiverse. I thank our younger son James for his patience and the wise words he has sometimes offered; we will fish again together in the Spring. I thank Stephen Windwalker of Kindle Nation Daily, who at my request joined our effort in California for several very intense days, while various very significant technical issues involving the publication process were ironed out; Stephen is the author of a leading book on new publishing technologies. I note that the incredible reliability of the Amazon system allowed us to receive research book orders, for one example, on only one day's notice, was a great help. Finally, I thank our daughter Paige Lehmann again for the cover art, which, though gathered together on just a few hours' notice, yet very well fit the requirements supplied.

Harry V. Lehmann
November 6, 2009

Prologue

Nobody knows for sure whether the Large Hadron Collider will destroy the Earth. Therefore, this book does not contain a "the sky is falling" claim that the implosion of the Earth into an artificially created Black Hole will occur.

Rather, two narrow points will be made; points which, if proven, compel a third: 1) The scientific possibility exists that the Large Hadron Collider (LHC), a huge physics experiment, owned by The European Organization for Nuclear Research (CERN), which is soon to start under at the Franco-Swiss Border, may destroy the Earth, and: 2) It is not possible (the main point here) for the scientists running the experiment to reliably assess the risk of a worst case event, so that: 3) Because of the known risk of planetary catastrophe, and that the impossibility of its occurrence being accurately ruled out, the LHC must be stopped.

That the possibility of the creation of a black hole exists, and that it could conceivably destroy the Earth, can in fairness be regarded as proven, *because this point has been conceded and even advocated by physicists supporting the project*, such as famed Oxford physicist Sir Martin Rees, who, in his book *Our Final Hour* quotes a study by three CERN theorists as concluding that if the LHC were to run for ten years, there would be a "one in fifty million" chance of destroying the Earth[1]. This was a CERN-collaborated study, telling us that CERN's activities are safe.

As will be seen, Dr. Rees' frequently quoted "statistic" was not actually a risk analysis based upon the LHC mechanism, although it is nonetheless widely relied upon in defense of the project, but rather it was a result from CERN *cosmology* theorists who published on the catastrophic potentials in 1999, as fully disclosed at page 124 of *Our Final Hour*, but as often incompletely credited on The Internet. In his stunning Prologue to *Our Final Hour*, Dr. Rees also describes what is at stake, the possibility, stated as exceedingly unlikely, that atom smashing colliders could start a chain reaction dissolving the Earth into nothingness, and that a tear in spacetime could then propagate from the Earth's former position, at light speed, a long shot Doomsday worst case, eventually engulfing our universe.[2] *Our Final Century* is available on *Kindle*™, the book is highly recommended for many reasons, and Dr. Rees' specific dialogue on the possible rent in spacetime is in his Prologue, readily viewable at the Amazon page for his book.

The stakes are thus higher than any previously faced by humankind: Perhaps some interstellar civilizations have passed "The Hadron Test" and others have not. Which shall we be? There have never been higher stakes. Yet, remarkably, we have not yet found a *truly independent experiment-specific study of the risk that world wide catastrophe might result from the operation of the LHC experiment.* For example, Jaffe, Busza and Wilczek, very highly qualified physicists, note in their paper *Review of speculative "disaster scenarios" at RHIC* that:

> "To our knowledge, possible catastrophic consequences of strangelet formation have not been studied in detail before.(fn) Although the underlying theory (quantum chromodynamics, or QCD) is fully established, our ability to use it to predict complex phenomena is imperfect."[3]

The Jaffe, Busza and Wilczek study, the most comprehensive and deeply qualified study of potential catastrophic risk from colliders, was not focused on the LHC, but on the RHIC, a different machine, and limited to strangelets, only, a particle which might not even exist. They said, see below, that the probability of a worst case event was higher at CERN than at RHIC, the collider which was their direct focus of their study, which study is nevertheless sometimes referenced in support of the LHC.

As is developed with citation in the body of this book, the most frequently cited "statistical" comment on such risk of catastrophe (one in fifty million, for which Dr. Rees is most frequently cited) was derived by these three employee theorists of CERN (Dar, DeRujula and Heinz) on the basis of *cosmological assumptions* regarding the estimated volume of the Earth, the estimated age of the universe, and the belief, stated as an observation, that such an event (in this case strangelet accretion) has not yet occurred, projecting therefrom that such a future event is extremely slight. So far as known to the undersigned, the possibility that many of the black holes which circle stars throughout the galaxies may have resulted not from "natural" phenomena, but from civilizations who failed what I call here "The Hadron Test" was not taken into account. A famed physicist, whom I count as a friend, and whom out of sympathy for his position I elect not to name, recently, on this point, stated to me via email that:

> "On another matter, the probability estimate of 1 in 50 million seems wrong to me. There appear to be lots of black holes out there, and regular supernovas, and we don't really understand why these things happen (there are models, but of course they are speculative and non-testable). Could some of these major celestial events have been caused by advanced civilizations that overstepped their knowledge and accidentally blew up their planets or solar systems?

I don't see why not. Where we are headed now might, among "star people" who survived their childhood, be known to be intensely stupid. Perhaps "they" care about a small planet on the edge of a rather common galaxy. Perhaps not."

I am not a particle physicist. I have carefully read the December 16, 1999 study by Arnon Dar, A., De Rujula, and Ulrich Heinz. I view them with great respect, and felt, from my keyhole, that their work was deeply orderly, thoughtful, and, certainly at a conscious level, sincere. I felt that their analysis of the strangelet possibility, rooted in cosmological odds (and thus, in addition, subject to the problem of Gaussian outliers, infra) was an elaborate way of stating what could also be a more directly calculation based upon the numerical relationship between the age of the universe, and a proposed ten year life cycle for the RHIC. In reading the A. Dar, et al, paper, I recognized that humility was in order for me; I felt that it was a beautiful paper. So, it is with very great humility that I mention the following comments concerning that study, so widely relied upon as a risk yardstick for the LHC.

1) The energies at the RHIC, as stated in the report, differ greatly from the LHC, and *linear extrapolations based on orders of magnitude* may not provide the level of predictive value sought. Simply stated, an "apples and oranges" problem. 2) The study is solely concerned with strangelets, not black holes, so that the study does not claim predictive value for the black hole issue set (even though the resulting Rees figure is very widely applied by various persons in defense of LHC safety in toto). 3) As above stated, and derivative from the second point, the study does not take into account the possibility, which has been ruled out by no one, that some detected black holes may in fact have arisen as a result of artificial manipulation. 4) Attention is respectfully noted to the closing paragraph under section 4 of the study, which includes: "It could be agnostically argued that, since the process of accretion onto a strangelet is surely difficult to model with confidence, our use of Eq.(7) is suspicious, and we should only abstract from it that the accretion times of different objects may be in proportion to the cubic root of their masses." This element, which appears to be an analytical pivot point, deserves brief comment. I do not concede it to be absolutely known that the accretion times for matter gathering to a strangelet, and matter, in comparison, gathering to a black hole, operate on the same temporal rate regime. In fact, some recent work questions the very existence of strangelets at all. Therefore, extrapolating to black hole accretion rates from this study is beyond both the data and its treatment, and therefore the study, despite the Internet flack, cannot be relied upon, even if it is good for strangelets, as a yardstick for black hole accretion rates, which in turn, thereby, shows limitation of the utility of application of the

study to the overall collection of risks presented by the LHC. However, further, as will be noted with specific citation later "black holes have profound similarities in the manner in which they accrete matter, regardless of their mass (M. I. T., Miller et all, supra, Chapter 9), and this later finding has application to the confidence that I, at least, have concerning strangelet accretion rates, as well, let alone black holes.

This possibility, that other some black holes in our universe may have resulted from failure of what I call here "The Hadron Test"(in that some black holes were not natural events, but resulted from arrogance), has not even been contemplated as a possibility in any of the studies of potential collider risk that I have so far encountered. Since the seed pearl, if you will, of a black hole cannot be known, it is true that exacting data based upon some percentile of artificiality cannot be used with any confidence in calculation. However, the fact, it is a fact, that artificiality of causation is within the realm of the possible, and, given numerosity, and what we see now on Earth, it is even, I submit, more likely than not that some level of artificiality has been a factor in black hole establishment. The fact of the possibility, which has not been ruled out, undercuts the reliability, potentially be orders of magnitude, of any Gaussian calculation process which seeks to base conclusions on an assumption of "all natural" phenomena, since the non-artificiality of some segment of the black holes remains unknown.

CERN, despite the prior findings of its own theorists that catastrophic destruction of the Earth is at least possible, though extremely unlikely, currently takes the contrary position on its own website, from a later study *by participants in its own venture*, that the chances of an Earth destroying catastrophe are "zero."

The Jaffe, Busza and Wilczek study, while focused not on the LHC, but rather another collider, called the RHIC, mentioned CERN in noting that: "Indeed, we believe the probability of a dangerous event, though still immeasurably small, is greater at the AGS or CERN energies than at RHIC."[4] Respectfully to all concerned, the obvious logic problem is that the study concluded that the "probability of a dangerous event" was "immeasurably small" and yet the odds of catastrophe were *greater* at CERN. Respectfully to all concerned, the concepts of "immeasurable" and "higher probability" cannot together describe the same phenomenon: If something is truly immeasurable, it cannot in logic be more or less probable than some other "immeasurable" event. This cuts to a major core issue presented to the readers of *No Canary*, namely that, ultimately boiled down, many allegedly "scientific" positions turn out to be zealously defended on the basis of emotion and hunch, in the utter absence (i.e., String Theory, infra) of empirical proof.

Full citations to the relevant studies will be provided in the later chapters; the bottom line for now is that the two major scientific risk analysis papers directly dealing with the question of immolation of the Earth by CERN's projects were written or participated in by CERN, or persons with CERN affiliations, which raises concern about the independence of the analytical process, and the third, and most comprehensive study in this issue area, was about the less risky (by the study's own words) type of particle accelerator, the RHIC.

The position of the scientists acknowledging the risk has been that, while the possibility of a worst case outcome exists, the chances of catastrophe are so remote as to be unimportant. This is the position of ever-popular physics author Dr. Brian Greene; that the *opinions* indicating against a worst case are so substantial that the project should proceed.

But to what end? Is there *anything* so valuable to the knowledge base of a sane person that taking *any* chance of killing all the Earth is justified? This book votes "no."

For reasons stated later in this work, with citation to relevant authorities, empirical factors, *including the empirical fact of lack of consensus on basic paradigms within the physics community on basic principles underlying the current risk analysis,* show us that current publicly visible science *cannot reliably gauge the chances of a catastrophic result.*

Finally, a moral question of literally astronomical dimensions will be brought forth in later chapters, involving the "Many Worlds" theory of the "multiverse." During the long study which resulted in this book, I learned to my surprise that the concept of a vast number of companion dimensions, each just like our own, with only, as to each, slight differences, is now widely accepted and taught as part of the curricula at major universities in which upper division and graduate level physics is taught, including Princeton (a Princeton physicist was prominently featured in the PBS special on Hugh Everett III).

In Chapter 19, we will, first, authenticate that this seemingly strange idea of "parallel worlds" has a substantial, credible, scientific following in the physics departments of major universities, and then, secondly, that this presents a vast moral dilemma, for how could it be moral for we, on the Earth as we know it, to hazard an extreme long shot of our own destruction, if we knew, or had solid reason to believe, that other worlds, Many Worlds, would for sure suffer destruction as a result of our own hubris?

Chapter 1

They've Blown It Up Once Already

On May 27, 2008, I submitted an amateur paper (the May 16, 2008 predecessor to this book) to *Review of Modern Physics*, which is the great old mainstay publication of the physics community in the United States. That initial paper, 27 pages in length, with 78 endnote citations, observed that:

> "As used here, the phrase "outlier outcomes" refers to the results outside the paradigms of the people responsible for estimating possible bad outcomes, as in, "I didn't know it was loaded." This might be important in the collider context because there is no miner's canary available to give early warning of outlier results: Outlier events, are outside of the expected, and may be sudden, or may be subtle and unnoticed, and yet ultimately of great consequence."[5]

After receiving the inevitable yet cordial denial letter from the *Review of Modern Physics*, I sent a brief follow up. I felt that my position on trans-universal entanglement in the multiverse, as a consequence of Big Bang "pre-matter" proximity was perhaps a new contribution of conceivable merit. Most importantly, this had been my best shot drawing attention to a moral dilemma which I believe stems from the modern physics belief, very widely held, in an infinite and expanding number of parallel universes:

> "My goals in submission of the essay to your publication were essentially humanitarian. The Gaussian based analysis of risk engaged in by the LHC does not provide an utterly complete assurance of safety, as is clear for many reasons stated in the essay, including Taleb and Gleick. Usually this does not matter, most experimentation involves some risk. However, in this instance, the risks, prima facie, potentially involve mass destruction. I am not an advocate of such a worst case outcome, but I am an advocate that it has not been ruled out."

The remainder of the letter was primarily concerned with moral and practical issues relating to the Many Worlds hypothesis, including a discussion of trans-universal entanglement, and those discussions will be rejoined commencing in Chapter 19. For now, it is sufficient to note the hard proof that, long prior to September of 2008, it was possible to predict that the LHC might be plagued by outlier events, not contemplated by its makers.

In addition to the above submission, our little team then sent mail merged letters, with a summary of central points, to sixty particle physics scientists, a majority of whom would soon be together at a then-upcoming conference; text portions from that letter are provided in Chapter 22.

Like a mouse from the grass, squeaking in fright at the noise of an oncoming freight train in the dusk, these were my feeble first attempts at slowing this project. This was writing squeezed into hours stolen from my intense active civil litigation caseload, which allowed only those few squeaks from the weeds in 2008. As it turned out, CERN had squeaks of its own in its perfect, "zero risk" machine; an accident two months later shut down the LHC.

On September 10, 2008 the LHC was started up for its first trial run, with protons being pumped in only one direction. Many in the public incorrectly believe that the LHC was tested successfully. That is not true. First, the protons were sent in only one direction, so that the collisions which are the goal of the project never took place. Secondly, without warning, critical elements of the LHC exploded.

The explosion at the LHC caused such severe damage that a full power re-start of the repaired machine may not occur until 2010, though lower power levels of operation are expected before the end of 2009. The originally submitted paper had also closed with the warning that:

> "The cascade from contemplation of multiversal entanglement is clear. If there is a "one in a million" shot that mini black holes or strangelets could cause the loss of the Earth, and such phenomena actually takes place in one or more of the vast number of parallel universes in the multiverse, it must necessarily follow that there is the potential for cross-multiversal entanglement, leading, at least eventually, to affect this universe.
>
> The above is either a legitimate point, or it is not. If it is a legitimate point, then the LHC should not be ignited until the possibility of multiverse entanglement has been fully vetted. In addition, the experiment is based on the assumption of light speed as an absolute, a well established concept, and yet a construct challenged by entanglement, at least. Should some combination of factors result in marginally higher velocities than anticipated, outcomes may be even more complex than it now appears possible."[6]

The point focused upon here is that CERN was not then, and is not now, all-knowing: The accident at the LHC was a practical demonstration that it is impossible for "bell curve" Gaussian probability analysis to accurately assess the likelihood of a worst case event at the LHC. The fact that a severe unplanned explosion took place at the LHC lays waste to all of CERN's persistent claims that these LHC scientists can accurately forecast LHC project risks; in fact, *as history has now empirically proven, they cannot.*

Chapter 2

The Hadron Test

This chapter presents two suggestions pertaining to the LHC: 1) The Hadron Test, and: 2) The Hadron Rule.

We ran Amazon searches for the terms "black hole" and "black holes" and came up with a total of more than *sixty thousand* search results for each term, both the singular, and the plural. Many scientists have studied them, many have written of them, and very obviously the educated public continues to buy the books.

Black holes are structures found, first through mathematical conclusion, and then, as instrumentation improved, through modern astronomy technique, which, result from burned out stars, the remaining mass of which was condensed through the gravity of remaining mass, into an object with gravity so strong that nothing can escape, not even light.

Despite all of the books, and all of the very well documented reasons why black holes come into existence, one startling additional concept is seldom if ever mentioned, and yet it has never been scientifically ruled out: That some black holes may have occurred from the conduct of beings on planets which formerly existed in their place. There is nothing to rule this out, and there is considerable data to support the possibility. As my physicist friend suggested: "Could some of these major celestial events have been caused by advanced civilizations that overstepped their knowledge and accidentally blew up their planets or solar systems? I don't see why not."

With stated odds like "1 in 50 million" we don't even have to go to the ethereal Many Worlds construction of physics to find new lyrics to *Paint It Black*: We have billions of stars in this galaxy, and billions of galaxies in this universe, their number still far beyond sure count, and since Earthly experience teaches us that hubris appears a universal constant, odds are, some worlds became toast long before breakfast was served. Don't let it happen here.

There is a very old saying in the Products Liability area of the plaintiffs' advocacy, where I have practiced for more than three decades, concerning the possibility that unique evidence may be "inadvertently" destroyed in testing while under the control of experts hired by the defendant corporation. In a humorous tone, yet earnest in intent, a seasoned lawyer may say, perhaps in a Boris Karloff tone to his or her younger associate, as the associate leaves to monitor an expert's test: Don't let them try the sulphuric acid test." Of course, say the defense counsel, nobody *meant* to damage the sample, "shit happens."

The discussion of destructive testing involves a duty of counsel, commonly encountered where there are a small number of evidentiary samples in a very large case: It is incumbent upon counsel to monitor, before the test, its scope, method, means, time limitations, equipment, and risks. We use the joke to teach a deadly serious point.

With the potentially fatal LHC experiment, and with other tests in the future, the question we should all focus on is: ***Should we ever risk our Earth, our Home Planet*** ? This gives a new twist to the popular [if new to the U. S.] term: "Homeland Security."

How could we have gotten to this place, that we are contemplating an experiment, and have spent billions on it, where there is credible science indicating even the barest *possibility* that it might destroy our Earth, with the odds against that tragic outcome being a mere matter of *opinion,* however lofty? The answer to this question is not found in the laboratory equipment, but in the hearts, minds, and institutions of those who have found a place to thrive in tax-supported science.

In 1998, my contingency fee tort practice, which helping many people who had gone through terrible tragedies, started to take a heavy toll on me. So, I took a break from that focus, and limited my products liability and personal injury caseload to a select few major cases where I felt my involvement would help greatly. I adopted a new business model, where, for the first time, I started taking "hourly" litigation work, for private plaintiffs and defendants, who by their circumstances has been forced to Court.

In those intervening years, before returning to a difficult quadriplegia case in 2007, I seriously avoided injured plaintiffs, and focused on public interest legal work, and took hourly litigation work, ordinary suburban litigation advocacy, to support the work in the larger cases. For about ten years I offered my services on an hourly basis to people who had no choice but to go to Court, often people who had been sued and had no coverage for the particular loss, but also several times for people who had to file suit because of persistent unreasonable and damaging conduct.

These "no choice left but Court" hourly litigation cases had one thing in common, to my surprise: the cases I personally encountered always involved narcissists. A neighbor doesn't like his downhill neighbor's trees, so he poisons them. A neighbor want so protect his own land from down hill sheet flow of water, so he constructs a secret pipe system to concentrate the flow and divert it upon his neighbor through a hidden pipe. A sister has hated her brother since his birth, so that now, when they are elderly, she forces him through expensive litigation with allegations which are later shown to be without merit, but at great cost, all to cause his pain. I had expected that angry people in this end of the work, which I had so seldom visited, but what surprised me was the uniformity of the self-absorption.

Perhaps you would agree that there has been a tendency in modern culture, you'd be blind to miss it, towards the satisfaction of one's own priorities, and, compared for example to the humanism of "the sixties" a reduction, at least in some quarters, for concern about the impact of one's activities upon our own species and others. Factory farming can be an example of this, though not always the case. Factory ships raping the ocean with gill netting of all species, killing dolphins along the way, in the interest of profit is another example. Coal must have its very essential place in our economy, but mountaintop strip mining, with "valley fill" certainly has placed a burden on the future that won't be felt by all those enjoying their profits in the present. I don't wish to make this political, and so will restrain further comment, but we can agree, can't we, that there is a tendency we have encountered, greater of late, to put one's own interests first?

So with that in mind, how much does the Higgs Boson mean to you ? Is finding one as valuable as your child? Or your soul, if you so believe ? It is not important in my life, for I cherish the miracle of life and consciousness as it is, every day, every minute. Yet, to many others, including very powerful institutions, it is *everything, and that devotion is obsessional.* We can tell for sure that it means "everything" to these folks, because they are willing to risk *everything* to find it.

My friend Bill M. recently asked me what all organizations had in common. I was providing a certain voluntary service to a great organization that we both respect, and that discussion led to his rhetorical question. I said that I wasn't sure what he meant, and waited.

Bill said; "Harry, all organizations are the same. Most start out with the best of intentions. Then, over time, they gain membership, and customs, and ways of doing things. Eventually, their goals will evolve to an extent which would be unrecognizable to their founders. But, however that goals part works out, they all want to do two things, grow and survive."

Later we will look at what the Longshoreman Philosopher Eric Hoffer, and other students of society, have had to say about this point. But this much must be admitted, personal satisfaction in the work, and membership in the institution, is a powerful force of reward for most institutions, including CERN.

So, who is to draw the line on experimentation, and where? In the very last of this, good scientific survey work will be cited which indicates that, in addition to any romantic or individualistic goals,

13

democracy is a worthy way of analyzing risk, because larger groups generally tend to have a higher degree of predictive accuracy. Yet: Is there some baseline principle upon which we can agree?

To pass The Hadron Test, we should set this boundary: <u>No scientific experiment shall ever be lawfully allowed which risks our very Earth</u>. This is what I posit to you as a logical limit, and there should at least be this limit, on the "scientific" activities of mankind. No scientist, or group of scientists, should ever be allowed to undertake, any experiment, of any sort, which risks the very existence of our home world, to any extent. This is what I call "The Hadron Rule."

Chapter 3

A Lunch With Charisma

Twice in recent years, famed trial lawyer Gerry Spence of Wyoming, has spoken at San Francisco's Commonwealth Club about books that he has written. Commonwealth is an outstanding organization, open to all, which provides thought provoking speakers on many and diverse subjects.

Gerry writes important books. I have heard him several times speak truths that others fear to mention, speaking with studied eloquence about the decline of our liberties, which are being washed away like the bleaching out of old jeans.

Gerry's daughter Kerry is a family friend of many years and, for these last couple of Commonwealth visits, I've hosted lunch after the speeches. The incident mentioned next took place at San Francisco's historic Taddich Grill.

A member of my small staff had called and arranged a table. After leaving The Commonwealth Club, we walked together along the streets of San Francisco towards the restaurant, hoping that Gerry's greatness would rub off by proximity, and we entered the narrow front, on one of the short sides of this long rectangle of a very classic restaurant.

As our group entered Taddich Grill, I saw to the left of our path that a private area had been set aside, a nice paneled room with a table set for ten or so. It fit our group like a glove, and was reserved and empty at a time when the rest of the place was packed. Surely, I thought, this had been reserved for our party.

Gerry, a walking mountain of karma, was caught up in a conversation, and hadn't noticed the reserved room. He plowed ahead like an icebreaker through Spring ice, with the rest of us in tow, trudging in a file through that long rectangle of a restaurant, to the far back, then up a few stairs, and then down a little hall, still moving towards the back of the building, until we finally hit a dead end, after which we all finally turned around.

At that point I finally spoke up, and so we walked back through the long axis of the restaurant, after our accidental tour, to that little paneled room by the front, and sat down for a great lunch.

I had known, as we passed that paneled room and started walking towards the back, that we were going the wrong way. I had seen the situation from the very start, but I had remained a silent walker.

I hadn't said, "hey you guys, hold on, look back here." Instead, in shyness (just socially shy, not shy in trial work), I had gone along with the crowd into the dead end at the back of the place. On that particular day, I didn't have the social "guts" to speak up: I "went along to get along." The stakes, back then, were low.

Now, with the world at stake, I still struggle, writing this, to overcome my intense embarrassment at claiming, as a mere lawyer, that preeminent scientists are wrong about risk assessment. My struggle has not been finding data supporting my position, of which there is a great deal, but to grasp guts enough to face society by writing this small signal flare about the unrecognized risks of the Large Hadron Collider experiment.

The interior struggle which I am experiencing comes *from the fear of being shunned for saying something against the tide of consented reality*. In this increasingly Global and Internet society, fear increasingly breeds silence, even from those who should know better. Despite all of our claims about individual liberty and free will, it yet remains that we historically tend to be pack animals, following Alpha.

As there was once an Iron Curtain, as communication increases, with its commensurate decline in individuality, there now grows The Wall Of Consented Perceptions, this is the Wall of socially defined ideas which we share with those whom we admire, those whom we fear and those whom we love.

There are many benefits to global consensus, notably the experimental hypothesis that war will thereby be reduced. Surely, we, as civil people, must share a consensus as to what is right and what is wrong, in addition to the lesser but essential laws about what is prohibited, and what is not. Yet, as the historian Paul Johnson illustrates repeatedly in his history, *Modern Times*, there always remains the "law of unintended consequences" which the modern military sometimes term "collateral damage."[7]

The reasons for the avoidance of war were brought home to planetary leadership by the First World War, where the British alone lost 900,000 people in trench warfare; WW II showed us the risk that we could bomb ourselves out of existence. So, there is value in learning to think together. Yet, all things in moderation: The problem that we increasingly confront as a world people is that an excess of managed and collective thought, for fun I reference *The Borg*[8] to illustrate, may so exclude individual epiphany; empirical risks of major catastrophe may be ignored as we collectively sit around the world campfire, singing "Kumbaya."

It has been found that the larger the group, in our Internet age, the larger the preclusion of inventive thought; so says a recent article from

Science[9]. According to social scientist Viktor Mayer-Schönberger of the National University of Singapore, software developers are now closely linked to project-specific networks. Mayer-Schönberger in essence reports that excessive numbers of connected participants on a project will reduce diversity of thought and restrain radical solutions from being vocalized. The report suggests that the path to innovation may be straighter if the number of participants in a project is kept low enough that fear will not restrain creativity. This article from *Science* was also reported upon *in New Scientist.* [10]

We fear to speak "out of line" and, the larger the group, apparently, the greater that fear. As social vision gets both larger, and congeals on a particular quested thought pattern, the fear of new ideas commences, in an expanding gradient, to also push out sensible complaint.

At that lunch with America's most famous trial lawyer, I wasn't willing to chance the embarrassment of speaking out in front of the group, as we walked in that very public place, even though I clearly saw the dead end coming. This fear of social blunder dominated the moment, even though I am in my fourth decade as a successful trial lawyer. Yet, in Gerry's shadow, in shy fear of social blunder, I remained quiet as we walked into a box canyon of a hall. Most of us do that. Scientists, as we shall see, do it constantly.

In this Intellectual Tribalism, we fear those with different positions, and we band together as our ancestors did, doubting and fearing all who may think or look differently than we do. Yet sometimes terrible historical events, or vastly silly ones, have happened when large segments of the population, or just an influential few, have failed, out of fear and conditioning, to speak up against foolishness, or even atrocity. When the stakes are truly high, and the moral compass clear, we owe it to our fellow people to speak up. And yet, history has shown, often we do not.

Chapter 4
Do Lemmings Dream of Parachutes?

"A man hears what he wants to hear, and disregards the rest."
- Paul Simon

The term "bubble" for describing runaway belief in the profitability of a given market, has entered the vocabulary of the general population. The term's popularity accelerated when millions of people were burned in recent years by such runaway beliefs. Recent examples include the "dot com" bubble, the "real estate bubble" and most recently, as credit default swaps, and similar betting motifs have dominated the headlines, by the "finance bubble." These bubbles share characteristic stages of development.

The first stage of a bubble is enchantment; a small group of people become enchanted with an ideal vision or system, such as a perceived probability of profit, or a new fashion trend, or something new in music. Sometimes the initial enchantment is both earned and deserved. For example, by the time The Beatles reached the United States, they were already great, through intense prolonged practice together on stage. Malcolm Gladwell, in his book *Outliers*, asserts that ten thousand hours of dedicated practice is necessary to true expert mastery of a skill. In that context, he reports:

> "The first interesting thing about the Beatles for our purposes is how long they had already been together by the time they reached the United States. Lennon and McCartney first started playing together in 1957, seven years prior to landing in America."[11]

Commenting on the band's time in Hamburg, Gladwell further reports:

> "The Beatles ended up traveling to Hamburg five times between 1960 and the end of 1962. On the first trip, they played 106 nights, five or more hours a night. On their second trip, they played 92 times. On their third trip, they played 48 times, for a total of 172 hours on stage. The last two Hamburg gigs, in November and December of 1962, involved another 90 hours of performing. All told, they performed for 270 nights in just over a year and a half. By the time they had their first burst of success in 1964, in fact, they had performed live an estimated twelve hundred times. Do you know how extraordinary that is? Most bands don't perform twelve hundred times in their whole careers. The Hamburg crucible is one of the things that set the Beatles apart."[12]

So, the thing enchanted about – perhaps an aviation hero like Mr. Lindbergh, perhaps a science hero like Dr. Einstein (see Walter Isaacson's magnificent biography for an understanding of Einstein's personality heretofore missing[13]), or musicians like The Beatles, or a revolutionary machine like an Apple Mac™ – may be the result of such high skill that there truly is a magical seed pearl to the enchantment of a generation, or even a nation.

Stage two is the stampede stage: There is broadly experienced obsessive engagement with the heroic subject matter, which is, in turn, associated with short term bliss. Others seeking the same obsessive rapture join the stampede. Soon, ordinary analytical paradigms are abandoned, and those who dare to suggest any application of prior analytical methods are, at best, spoken of as "not with it." At worst, those who question the rush to judgment are burned at the stake, if not literally, at least in that their academic careers go up in smoke. In this instance it is inevitable, and yet also fair, that the Nazi experience in Germany be mentioned, as an example of runaway adoration of a particular system of thought.

The history of group rapture dates thousands of years, though surely accelerated by the advent of the printing press. Perhaps there may be a physiologic element common to these constituents of group rapture; perhaps enhanced self respect is accompanied by serotonin increase, crutched by the group. Dr. Peter D. Kramer, in *Listening to Prozac*, explains that a person being dominated by another will sustain a diminishment in serotonin concentration, while the person doing the dominating may at the same time get a lift in the same chemical[14]. Perhaps participation in the rapture of a Bubble resembles a gambler with a slot machine, getting a little serotonin or endorphin or similar thrill associated with the risk-reward cycle of the experience.

If enchantment is the first stage of a "bubble" perhaps we can agree that obsession often follows as the second, followed by the dangerous third stage of "stampede" and then finally, in the fourth "burst" stage, the lemmings go over the edge of the cliff.

In the inevitable fourth stage, the bubble is "burst" and all involved, including the wounded spectators, recognize that "something went wrong" and there is a blame festival amongst commentators to reach consensus on a label for the disaster.

Whatever the many roots of perceived "happiness from membership" it is nourished by adhesion to the popular line, sometimes also called the party line. We in human society often act like herd animals; each of us is fearful of rejection; many of us also using fear of rejection by others as a means of asserting control. And, though not a central aspect of this work, those of us in the herd often don't even notice the cowboys.

The tendency of humans to think in herds, and the recognition of management systems intended to inspire and manage "correct" herd thought has been treated by dozens of serious authors, from Vance Packard, in *The Hidden Persuaders*, to Jacques Ellul in *Propaganda, The Formation of Mens Attitudes,* to advocates of such control by "Controllers" in *Beyond Freedom And Dignity* by B. F. Skinner. The results of such attempts towards governmental control of humankind, always well intended from some perspective, is perhaps, as to the 20th Century, best covered by Paul Johnson, in his monumental and well documented history, *Modern Times.* Poetry, too, has dealt with our evolutionary history of fear: Jack London, in *Call of The Wild*, took our minds to the interior thoughts of his furry protagonist, Buck, as the giant Husky in London's poetry caught visions at a campfire in the arctic dark.

> "Sometimes as he crouched there, blinking dreamily at the flames, it seemed that the flames were of another fire, and that as he crouched there he saw another and different man....This other man was shorter of leg and longer of arm, with muscles that were stringy and knotty rather than rounded and swelling. The hair of this man was long and matted, and his head slanted back under it from the eyes. He uttered strange sounds, and seemed very much afraid of the darkness, into which he peered continually, clutching his hand, which hung midway between knee and foot, a stick with heavy stone made fast to the end. He was all but naked, a ragged and fire-scorched skin hanging part way down his back, but on his body there was much hair. In some places, across the chest and shoulders and down the outside of the thighs, it was matted into almost a thick fur. He did not stand erect, but with trunk inclined forward from the hips, on legs that bent at the knees. About his body there was a peculiar springiness, or resiliency, almost catlike, and a quick alertness as of one who lived in perpetual fear of things seen and unseen."[15]

From an evolutionary standpoint, fear of non-conforming thought is a survival mechanism; when we see those dressed differently, or accented differently, there is a tendency, often encouraged by those who use fear as a tool, towards interior denial of their very humanity. This denial of humanity, of course, is accentuated in times of conflict. Thus, during the Viet Nam war, the television news reports would tally the numbers of "VC killed today" as though they were sports scores. Civilization depends upon not allowing the course of our future to be dictated by whim, and yet, as noted in the study on the diminishment of thought quality in Internet grouping cited earlier, there is always the possibility that we may, all singing a chorus of comfort in the main lounge, still yet hit an iceberg in the dark.

Thus, commentators on both the political "left" and the political "right" tell us that public perception is shaped to a particular point of view. From the far Left, Noam Chomsky wrote in *Chronicles of Dissent*:

> "Our system works much differently and much more effectively [than the Communist system]. It's a privatized system of propaganda, including the media, journals of opinion and in general including the broad participation of the articulate intelligencia, the educated part of the population. The more articulate element of those groups, the ones who have access to the media, including the intellectual journals, and who essentially control the educational apparatus, should properly be referred to as a class of 'commissars.' That's their essential function: to design, propagate and create a system of doctrines and beliefs which will undermine independent thought and prevent understanding and analysis of institutional structures and their functions."[16]

Despite our claims of individual free will, some like Chomsky would aver that the decisions which we choose are all from our own conditioned list of "approved thoughts." Our choices are plucked from a stream of alternatives which is systematic in nature, and we feel great discomfort should an idea be presented which differs from the consensus reality in which we, as purportedly thinking individuals, have chosen to partake. This phenomena is non-partisan – having taken a look at what the radical iconoclast has to say, let's look at counter-point from a Conservative view.

Dr. Thomas Sowell, in *The Vision of the Anointed*, observes that there is a repetitive pattern also seen in those who seamlessly defend the virtues of government:

> "Despite the great variety of issues in a series of crusading movements among the intelligentsia during the twentieth century, several key elements have been common to most of them:
>
> Assertions of a great danger to the whole society, a danger to which the masses of people are oblivious.
>
> An urgent need for action to avoid impending catastrophe.
>
> A need for government to drastically curtail the dangerous behavior of the many, in response to the prescient conclusions of the few.
>
> A disdainful dismissal of arguments to the contrary as either uninformed, irresponsible, or motivated by unworthy purposes."[17]

In *Vision of The Anointed*, Dr. Sowell gives his version of the actual effects of the War On Poverty, launched by the Economic Opportunity Act of 1964, which was motivated by a stated desire to reduce dependency. President Kennedy favored the Act, as reported in the Congressional Quarterly of February 2, 1962: "The President stressed that the welfare program should be directed toward the prevention of dependence and the rehabilitation of current relief recipients."[18] However, despite the well laid plans of a well intentioned Establishment, Dr. Sowell notes that: "The number of people receiving public assistance more than doubled from 1960 to 1977. The dollar value of public housing rose nearly five-fold in a decade, and the amount spent on food stamps rose more than ten-fold."[19]

Academics who had favored the program (disclosure; the author received federally guaranteed student loans in the 1960's, and in 2008 supported and voted for Mr. Obama) rationalized that things would have been worse had they not been in place, as Dr. Sowell describes:

> "Finally, it was asserted that things would have been worse were it not for these programs. 'The question is not what the bottom line is today–with poverty up–but where would we be if we didn't have these programs in place?' asked Professor Sheldon Danziger, director of the University of Wisconsin's Institute for Research on Poverty, 'I think we'd have poverty rates over 25 percent.' Even though poverty and dependency had been going down for years before the 'war on poverty,' began, Professor Danziger chose to assert that poverty rates would have gone up. There is no possible reply to these 'heads-I-win-tails-you-lose assertions except to note that they would justify any policy on any subject anywhere, regardless of its empirically observed consequences."[20]

No importance is attached here to which side was correct in the above *"he said – she said"* argument about entitlement problems. I'm happy to assume that the truth is someplace in between. The important point, brought out by Dr. Chomsky, more on the Left, and Dr. Sowell from a Conservative perspective, is that advocated positions by educated people in professional settings often adhere to a scripted viewpoint, adhered to out of a psychological need to belong and prosper, and *not as the result of empirical analysis.*

This willingness to choose an "approved" thought process, over an "empirical" thought process, appears in all levels of profession, including particle physics.

Chapter 5

Tulips and Poses

Nassim Nicolas Taleb, in his groundbreaking *The Black Swan: The Impact of the Highly Improbable* quotes the 17th century philosopher Simon Foucher, saying: "We are dogma-prone from our mothers' wombs."[21] We filter out as unacceptable ideas and statements which are not consistent with the "comfort level" of our world view. This tendency of humans to "group think" is visible not only in small scale, or "micro" social interactions such as we all have experienced or witnessed: History shows that we humans, for all of our claims about individuality, at a large scale, governmental and "macro" level, tend to run in packs, sometimes with no more moral concern than wild dogs.

Taleb's work also illustrates how, in our modern, media saturated age, an attractive idea can be so overwhelmingly attractive that competing intellectual ideas are utterly ignored by the body politic. Thus, Taleb notes the words of Captain E. J. Smith, later Captain of The RMS Titanic, as spoken in 1907:

> "But in all my experience, I have never been in any accident ... of any sort worth speaking about. I have seen but one vessel in distress in all my years at sea. I never saw a wreck and never have been wrecked nor was I ever in any predicament that threatened to end in disaster of any sort."[22]

In the worst cases, this tendency to group identity has been a root tool of those seeking war, where propaganda is used for their own ends. The examples of this are legion, and not always part of the popular history which results from "winner's justice." Short of war, yet very far reaching in consequence, this human tendency of individuals, and of the institutions in which they congregate, accepting only "group approved" ideas, even where rational scientific method might lead elsewhere, has had effects not only in the economic fields and disciplines, but in every aspect of human academic endeavor, including the sciences. Yet it is in economic history where the most colorful, and perhaps least tragic ("it's only money") stories reside.

The year 1841 included the publication of the book *Extraordinary Popular Delusions and the Madness of Crowds,* by the British journalist Charles Mackay, still available, the earliest book I have so far found describing the event notoriously called "Tulipmania" or, as recently titled in distinctive title spelling by Mike Dash: *TULIPOMANIA - The Story of the World's Most Coveted Flower.* These books, and others, notably including the encyclopedic and profusely illustrated *Tulipmania: Money, Honor, and Knowledge in the*

25

Dutch Golden Age, by Anne Goldgar, of King's College London, tell the story, from differing perspectives of the "Tulip Bubble" which was a central aspect of the Dutch financial scene for roughly the years between 1624 and 1635. This "Tulip Bubble" is frequently referenced even today by those writing about the field of finance, as a cautionary tale about crowds of people rushing to a destination of financial disaster, a captivating story about the herd tendencies of human belief.

Mark Frankel of *Business Week* presciently titled his April 2000 review of Mike Dash's Tulipomania; "When The Bubble Burst" and described the relevance of tulips to silicon at that point in financial history:

> "Long before anyone ever heard of Qualcomm, CMGI, Cisco Systems, or other high-tech stocks that have soared during the current bull market, there was Semper Augustus. Both more prosaic and more sublime than any stock or bond, it was a tulip of extraordinary beauty, its midnight-blue petals topped by a band of pure white and accented by crimson flares. To denizens of 17th Century Holland, little was as desirable. Around 1624, the Amsterdam man who owned the only dozen specimens was offered 3,000 guilders for one bulb. While there's no accurate way to render that in today's greenbacks, the sum was roughly equal to the annual income of a wealthy merchant."[23]

Mr. Frankel's article was exquisitely timed - the dot-com bubble had started to fill by frenzy in 1998, and there had been three years of incredible boom in Internet stocks. Reminiscent of the later "Real Estate Bubble" the fact that "what goes up, must come down" had been lost in emotive enthusiasm, culminating with the Nasdaq index reaching an intra-day peak of 5,132.52 on March 10, 2000, after which the decline commenced, and then accelerated, to the ruin of many, and Frankel's review chimed in the very next month.

Though hundreds of years before technology stocks, the Tulip Bubble was very similar in that its sails were filled with such storm winds of emotion that the staid investment practices of the 17th Century Dutch were cast aside. Thus, Mike Dash writes in his remarkable best seller that:

> "From the Autumn of 1635,then, the bulb trade changes fundamentally and forever. Ignoring the customs of the connoisseurs, increasing numbers of florists progressed from trading only tulips that they had in their possession to buying and selling flowers that were still in the ground. Bulbs then ceased to be the unit of exchange; now the only thing that changed hands was a promissory note – a scrap of paper giving details of the flower being sold and noting the date upon which the bulb would be lifted and available for collection.

"There were advantages to the new system. It certainly permitted trading to take place throughout the months of autumn, winter, and spring, and because the bulbs stayed where they were until lifting time no matter who their new owner was, it was very appealing to florists who had neither the skill nor the desire to cultivate bulbs themselves. . .

"The Dutch called this phase of the tulip craze the *windhandel,* which can be translated as "trading in the wind." It was a phrase rich in meaning. To a seaman, it meant the difficulties of navigating a ship steering close to the breeze. To stockbrokers it was a reminder that both the tulip traders' stock and their profits were so much paper in the wind. To the florists, however, *windhandel* mean trading pure and simple, unregulated and unconfined."

"It was this innovation that made the greatest excesses of the mania possible. The introduction of promissory notes did much more than make the tulip trade a business that could flourish all year round; it turned dealing into an exercise in speculation, and - because delivery was usually months away – it encouraged the sale and resale not so much of bulbs but the notes themselves."[24]

Sound familiar? Just substitute "credit default swaps" for the promissory notes on tulips, and we're off to the races. There are several good books on the Dutch tulip mania, but Mike Dash's one paragraph portrait of the first day of the crash tells the story of the first Tuesday in February, 1637 with poetic clarity:

"As was customary, an established member of the college began the day's trading by testing the state of the market; he offered a pound of Witte Croonen or Switsers for sale. The florist asked a fair price – 1,250 guilders – for the bulbs, and in the normal course of events he would have had several eager buyers. Slates and chalk would have been distributed, the tulips would have been knocked down to the highest bidder, and the rest of the day's trading would have continued in its usual frenzied way. On this day, however, there were no bidders for the bulbs at 1,250 guilders. The auctioneer offered them again, this time cutting the price to 1,100 guilders. Still, there was no interest. Desperately now, he offered his bulbs for a third time, dropping the price to a risible thousand guilders to the pound. Once again, there were no bids.

"It is easy to imagine the awkward silence that must have descended upon the group of florists hunched around their table in the tavern as this awful, farcical auction ran its course."[25]

The collapse of the tulip market had a domino effect on the more general Dutch economy, certain to have occurred, but difficult to objectively measure due to the level of records at the time. But, things returned to prosperous normalcy in a few years, and, as we say here on Earth, "C'est la vie." But that was just one market, in one country, in one corner of our vast and intricate globe. The Dutch got over it. Other examples of group think have not faired so well, as even a brief visit to the totalitarian excesses of the 20th century well illustrates. There are benefits to working together, but also dangers inherent in thinking together, and the stronger and more indelible the belief, the greater those dangers.

Chapter 6
Over The Cliff

Mike G., still my friend after almost fifty years, was the first serious surfer-kid to come to Novato High School. He arrived from Half Moon Bay when we were sophomores, in 1962. His timing was lucky. This was the era of surfing images and surfer songs. Songs by Jan and Dean like *Surf City*, and The Beach Boys with songs like *Surfin' Safari*, *California Girls*, and *Surfin' USA*. Use your PDA to download "Surfin' USA" and "Surf City" and you'll be there. It was a time of *enchantment*. Mike, having arrived with his own surfboard, way before that was common (9' 4", and not even a stringer), was impossibly popular.

It was a great time to be a youngster in California. We had, after all, California Girls. And beyond the imagery, there were unforgettable days in the salt water; for those of us from Marin, this was out at Bolinas, at The Patch, Brighton Street, or The Jetty, or The Channel. For those who truly stuck with it, it also meant Steamer's Lane, where I once in later years lost my board to the rocks, and very nearly my life.

In the pre-Hippie 60's, The Beach Boys were constantly on the radio and the radio was constantly on. On warm Saturday nights when we would cruise Fourth Street, in downtown San Rafael east to west and then back again ten miles south of Novato. Novato was very rural when I grew up, raising ducks in the back yard, but all that was changing by the mid-sixties.

The reason that I bring this all up has to do with Mike's two sweaters. In the Fall of 1963, in our Junior year, just as the leaves were turning brown at the first chill, Mike wore a fuzzy yellow V-necked sweater to the halls of Novato High. Such a sweater had not been seen before at Novato High. It was, after all, pastel. Within two weeks, virtually every boy at the school owned and wore a yellow fuzzy sweater. The school mascot at Novato is the Hornets, and between classes at that phase, with the boys milling and walking all in yellow, the name fit.

Six weeks later, Mike wore a blue cashmere V-neck, again previously a rarity at most; that was the next "must have" item; the school hallways looked like moving rivers of blue and yellow.

During the same period, my friend Rod M. wore a pair of black Converse tennis shoes to school. Before Rod wore them, wearing such a thing was a guarantee of ridicule. As soon as Rod, who was in a garage band with Bryce and Larry, wore those Converse shoes, they were the shoes to wear: Hundreds of boys started wearing black Converse tennis shoes to go with their pegged jeans; for a while the stores were out of the shoes. Everybody had to have pegged jeans and "acceptable" clothes.

In remembrance: Rod was the son of retired high altitude pilot Joe M., upon whom Chaing Kai Chek had pinned medals, and who in later years became my treasured and closest friend, Joe and I fished in deep waters out to Cordell Banks, out of Bodega, in which waters Joe won 38 pools for largest fish in his last active year fishing, in 1994. Joe was so tough he was at least once sent after black money gone south. Joe is gone now, my valiant, tough, best fishing buddy, and I mention him now purely in loving remembrance. Joe's military rank belied his actual rank as a civilian pilot, of sorts.

Those high school years, the sweaters, the Converses shoes, were my introduction to the fact that we, as a species, in music, in popular culture, in our willingness to believe tall tales if backed by forceful imagery, are Cultural Lemmings. Later, in politics, as a professional state-wide campaign director, I once saw preference for my candidate, go up 38 percent on college and university campuses in California, during the same five or so months that the program I had created and was managing was aimed and operational towards just such a result. This was in the early 1970's, when it was still possible to be a "Liberal Republican" (an endangered species now verging on extinction), and I had found the effectiveness disturbing in a way, which contributed to my long held determination since then to avoid the political trade.

Chapter 7
For My Skeptical Friends

If the Large Hadron Collider experiment may destroy our planet, perhaps my writing should be more like a zealot's, and less like a tired lawyer's. In fact, this isn't a favorite subject area for me; instead I feel compelled by conscience to pen this book: My interests are in other fields, where I hope some day to write. But I've studied the sciences for many years, and their interface with the law, because my practice area for three decades has been the civil advocacy of cases built on scientific proof.

I chose this course because there is less bullshit in engineering cases than those based on more ephemeral ideas. One can't, for example, truly understand the pressure differential problems inside a levee at flood, unless one has a passable understanding of hydrology principles, such as the relationship between water velocity and the ability to carry particulate matter. One can't, for example, truly understand the brake-failure taxiway crash of two expensive aircraft unless one understands Pascal's Principle, the basic principle of hydraulics.

So, we should reach for scientific truth, and also be able, of great importance here, to recognize scientific uncertainty. As we grapple with this, surely we can agree for sure on this much: <u>Any legitimate arguments relating to the LHC should be grounded in fact-based scientific reasoning</u>.

Thus, we can agree that if a "science" position isn't based upon scientifically replicable empirical data, then it shouldn't be relied upon where major issues of safety to life are concerned. Let that be a "no mumbo-jumbo" pact that you as the reader, and I, as the writer, share in this book

Andrew Newberg, M.D., is an assistant professor in the Department of Radiology in the Division of Nuclear Medicine, at the University of Pennsylvania. With Eugene d'Aquili, M. D., Ph.D., (who taught in the Department of Psychiatry at the University of Pennsylvania for twenty years), Newberg co-wrote a book called *Why God Won't Go Away*. In the book, he defines scientific thought nicely: **"Science concerns itself with that which can be weighed, counted, calculated, and measured, anything that can't be verified by objective observation simply can't be called scientific."**[26]

You and I agree on that, right? That question requires a "yes" or "no" answer. You either agree with me that a scientific approach must be based in fact, or you do not. If a belief isn't based on facts, then it is faith-based, which is fine for church (not said cynically) but I don't want my surgeon relying on the Bible for his operating manual.

Thus I invite being held accountable in this book to the same standard stated above: "Science concerns itself with that which can be weighed, counted, calculated, and measured, anything that can't be verified by objective observation simply can't be called scientific." Classically, the law is based on the very same sort of reasoning, though muddied by politics in recent years. There will be no faith-based arguments in that which follows.

Chapter 8

How We Measure Risk

You've heard the joke that "surgeons bury their mistakes." Lawyers are in a very different situation. We bury our successes. Because of the confidential nature of our work, some of the best work, done by talented lawyers, is never seen. Sometimes this is because the tactic itself is not seen by the opponent. Sometimes this is because of the subtly different reason that while the tactic is seen, the underlying system of reasoning behind the tactic remains obscured.

Because my work involves scientific proof; failures of aircraft systems, failures of Earthen dams, a long list, I joined an international association of Earth scientists, the American Geophysical Union, an excellent organization deserving of support. When attending meetings of this organization the academics involved often look quizzically at my convention badge, stating my professional affiliation. The looks would say, "what is a *lawyer* doing here, amongst us scientists??"

The most important contests in our lives are entrusted to the legal system for resolution. Therefore, lawyers are trained to reason in a scientific methodological way, training which sometimes sticks, and sometimes does not. Often these controversies involved scientific study; consider a medical malpractice case, for one example; to properly prosecute or defend the case, the underlying medical science must be understood by the lawyers and the Judge involved, which requires very deep study.

One of the main baseline ideas lawyers learn about is negligence. Negligence means pretty much the same thing in Court as it does in real life; somebody made a bad mistake, and somebody else was hurt as a result, or money or property lost or destroyed.

Since we, as a society, discourage civilian gun fights, the resort is to the legal gunslingers of the State and Federal courts. A plaintiff who seeks to prove a case based upon negligence has the burden of proving the facts which amounted to negligence on the part of the defendant.

In order to learn about negligence, law students study hundreds of appellate cases, and they are required to recite and analyze those cases, before their peers, when called upon without warning by the professor involved. At least that is the Socratic process that went on where I went to school.

Out of all the cases used in the teaching of the law of Torts (which includes negligence), none is more famous than Justice Cardozo's 1928 decision in *Palsgraf v. Long Island Rail Road,* 164 N. E. 654. The decision in Palsgraf is so well known that if you ask your lawyer buddy about it, he or she will still remember it, and probably recall the facts, and what the case stood for.

The main point of the Palsgraf case is that the level of *duty* that one person owes to another while engaged in an activity is "strictly correlative" with the risk involved in that activity. So, if a very grave harm may result from an activity (such as dynamite blasting, for example), then the level of care owed by the person doing that activity is very, very high.

From this simply stated example, we can all see that if the actual destruction of the Earth is at stake, then it would be required that "absolute" care be taken to protect us from that bad outcome. Even more than that, there are certain sorts of activity which are so very dangerous that we simply prohibit them; we call them "crimes" this is where the risk of an activity to others is so high, that the activity is prohibited.

For the reasons next stated, it is suggested that such a prohibitory outlook should be immediately adopted as to the LHC; that its operation should be stopped through every lawful means, since there are strong data indicating that: A) The destruction of the Earth may possibly result from the operation of the machine, and B) It is not possible for the operators of the machine to accurately determine the risk which is presented.

Chapter 9
Nothing But The Facts

An AP story of June 28, 2008 nicely summarized the competing positions about the LHC: Proponents of the project say that there is "zero" risk; other physicists state that there is "only" a "one in fifty million" chance that the Earth will be destroyed. Big lotteries have been won against longer odds.

No situation of this magnitude has been seen previously by mankind, due to the enormity of loss if the "experts" are wrong: Proponents of the project predict that it may result in the creation of "mini black holes." Other writers take the seemingly bizarre position that the creation of "strangelets" may cause conversion of all matter in the Earth into "strange matter" destroying life. Widely encountered defenses of the LHC are on the basis that the "mini-black holes" if created, will have "the energy of a mosquito" so that therefore concern is irrational and safety worries are routinely called "crackpot" by its defenders in the first paragraph of defense. Others, as noted on the CERN website, have expressed the fear of a runaway fusion reaction, which CERN says is impossible.

Since the result of any one of the above noted risks, dismissed by CERN, would be catastrophic for the Earth, proof that CERN's contentions about safety, as to even one such risk, should be enough to inspire the delay of LHC ignition until each such issue, including moral issues ("many worlds") has been independently and objectively vetted.

In that which follows, it is argued, with citation, that traditional Gaussian ("Bell Curve") risk assessment, core to the LHC risk analysis so far undertaken, has now been variously shown to be subject to severe error, in which regard the work of Taleb is valuable. Other risk factors to be surveyed here will include:

- MIT's Center for Space Research published a report by Miller on observations of black hole J1650-500, ten solar masses, contrasted with a ten million solar mass black hole in galaxy MCG-6-30-15; the stated conclusion is that; "black holes have profound similarities in the manner in which they accrete matter, regardless of their mass."[27]

- There is a relevant history of organized scientific catastrophic result, including famously The Titanic, the Gemini launchpad explosion, the explosion of Columbia, the loss of Challenger, and let us not forget the Mars Climate Orbiter. Then there was the SSC "hole" at Waxahachie, Texas, the first attempt at an

LHC, where scientifically predicted costs were overrun three-fold, from about $4.4 billion to a projected $11 billion, at which point Congress finally pulled the plug.

- The long history of scientific error is independently significant in this situation because "everything" is at stake.
- Chaos also presents unpredictability here.
- Most LHC scientists believe in String Theory, *for which there is no empirical proof,* see Peter Woit's book, *Not Even Wrong.*
- In the multiverse, a tiny chance here can mean catastrophic destruction means a lot in the affected alternative dimensional realm.

As will be developed in the chapters which follow, there are legitimate science-based reasons to seriously question the level of risk analysis at the LHC.

Chapter 10

No Canary In The Quanta

The Large Hadron Collider will hurl protons at one another through a seventeen mile long circular tunnel, until they crash into one another at more than 99.99+ percent of the speed of light. Controlled head-on sub-atomic collision velocities this fast have never been reached before in the whole history of mankind as we know that history.

If everything goes as planned (an unknown, since the scientists involved already blew up the machine on the first try), these unprecedented proton speeds, made possible by physically huge and incredibly powerful computer controlled electro-magnets, will allow the breaking down of matter in a way and to an extent never before accomplished.

Countless thousands of physics enthusiasts hope that the ejecta traces from these collisions will yield clues about the (as yet unknown) underlying nature of our existence itself, by re-creating the first moments of the Big Bang which started our universe, yep, *right here on Earth!*

This is called the most expensive experiment as yet undertaken, and for civilian science, that is true. My concern remains: *They don't call it the Big Bang for nothing.*

Last year, a sensitively composed story in the *New York Times* covered a lawsuit, filed in U. S. District Court in Hawaii (and since dismissed on the technical ground that the U. S. Government was not sufficiently involved to give rise to Federal jurisdiction) by a chap named Wagner, who has filed a couple of "doomsday" suits before, concerning smaller and less powerful colliders which certainly *appear to be* humming along just fine[28]. I use the phrase "appear to be" in direct reference to the multiverse problem set and the vast moral implications that Everett's work, as covered previously, has in this context.

Skepticism is always warranted. The LHC lawsuit, which is not a main focus here, sought to enjoin some American entities, Fermilab for one, from further support of the LHC project. Why? Well, here's the sticky part: Mr. Wagner and some other folks argue that messing around with the very starting point of the universe, which is the stated goal of the LHC, may not really be a bright idea.

The *New York Times* article on the lawsuit was sensitive to the central underlying issue; its author, Mr. Dennis Overbye (who seems to have stepped into James Gleick's giant shoes) wrote: "Although it sounds bizarre, the case touches on a serious issue that has bothered scholars and scientists in recent years – namely, how to estimate the risk of new groundbreaking experiments and who gets to decide whether to go ahead or not."[29]

In the 1970's, California for the first time adopted, as qualification for the practice of law, a requirement that lawyers pass an examination on the subject of Professional Responsibility (legal ethics). Prior to that time, it was believed, and not without cause, that ethical precepts were either obvious, or would "seep in" though exposure to the rest of the law, which, in its majesty of story, does tend to encompass the whole epic of human experience. That is a statement which will be best understood by readers who have needed a good lawyer in a tight spot. So, in preparation for this book, it was natural for me to search for some generally approved set of ethical guidelines for scientists. Though ethical standards can be found in some areas, notably medicine, I have yet to find of a Code of Scientific Responsibility. Why? Well, here we get to the question of moral relativism, and if one, I suppose, truly believes that "it's all relative" then a true Code of Ethics would be, well, constricting, I suppose. So, I can't use the Scientists Code of Ethics as a reference point here, because the scientists don't have one.

What they do have are talking points. Upon examination, many of those points can be shown as articles of faith.

Those defending the LHC have pointed out that the incredibly tiny size of any sub-atomic black holes, allegedly with no more energy than a mosquito, coupled with their predicted evaporation, means that there is no practical danger to the Earth. On the other hand, one of the scientists making this "mosquito" argument also pointed out that it was advantageous to have the tunnel underground, since the beams involved can cut through fifty meters of rock.

Exactly contrary to the "size doesn't matter" defense of the LHC, findings at MIT's Center for Space Research indicate that *the manner in which black holes accumulate matter is independent of their mass.* The MIT news office, in its release on the work of a team led by Jon Miller, reported observations of a "small" black hole named J1650-500, which is 26,000 light years from Earth, and has ten solar masses[30]. Miller's findings were discussed at a 2002 joint meeting of the American Physical Society and the High Energy Astrophysics Division of the American Astronomical Society, and these findings were derived from observations from the European Space Agency's XMM Newton X-Ray satellite. The work compared the findings regarding J1650-500 with findings regarding a 10-million-solar-mass black hole in galaxy

MCG-6-30-15. It concluded that the smaller black hole in our galaxy behaves nearly identically to the million times larger black hole 100 million light years away. The MIT release states that: "This indicates that black holes have profound similarities in the manner in which they accrete matter, regardless of their mass."[31]

The application here is obvious; a black hole of any size, including the subatomic variant, may possess the capacity to "accrete matter" at the same rate as any other black hole; from our perspective, a catastrophically rapid event.

There are scientific arguments against the supposed risk from black holes since some LHC proponents, such as Brian Greene, expressly state the possibility that black holes may results, the sometimes-stated "they won't happen" argument, can be summarily dismissed. More substantive is the argument that, due to "Hawkins radiation" (a theoretical construct), the LHC black holes will not be "stable (and therefore accrete matter) but will evaporate instantaneously. A response to that second argument is put forth in Chapter 25 of this book.

Chapter 11

Limitations of Gaussian Risk Analysis In Regard

To The Large Hadron Collider

Dr. Brian Greene, of superstring fame, was asked, in an exchange I witnessed, *if the LHC might spawn whole new universes.* This was in Marin County, California, at our Civic Center theater on March 6, 2006.

Dr. Greene said, "*yes*, it could, but we just don't know." He spoke with an excited Maestro's upward wave of the hands, in emotive enthusiasm in describing the upcoming super collider experiments at the European Center for Nuclear Research (CERN).

The questioner appeared perplexed, but the inherent question "should we?" was never addressed by Dr. Greene. The inherent issue of whether hubris was affecting scientific planning was either intentionally ignored or selectively not perceived.

In his enthusiastic comment, Dr. Greene, the most public figure in sub-atomic physics (for his elegance in getting his "String Theory" across to a popular audience), was joining with a huge chorus of other scientists caroling about the LHC, as though Santa were coming to town: "We just don't know and we're gonna find out, LHC is coming to town."

As will be seen, many leading physicists have, like Dr. Greene, truly retained their sense of child-like wonder. For example, Dr. Greene was interviewed in March of 2009, on the question of whether or not the LHC would create *new universes.* His sense of wonder remaining intact, Dr. Greene seeks that goal.

In the interview with Dr. Greene, taken on March 25, 2009, the point is made that an ultra tiny "seed" such as may possibly be created by the LHC, can, under the right circumstances, based on current calculations based in General Relativity, give rise to the "budding" *of an entire new universe.* The reader is referred to the entire podcast for full text[32], but the following quotations from Dr. Greene himself illustrate, in my view, the dangers now presented to we, the people of Earth, from the LHC:

> "There are conditions, which, according to the laws of General Relativity, the laws that Einstein wrote down a long time ago, well-tested, those laws tell us that, in this context, in the right energy density carried by the right substance, you will have repulsive gravity, which means that if you can build this little seed, this little nugget, in just the right way, it will, on its own, roughly speaking,

41

start to expand, grow, faster, faster and faster, sprouting, into a gigantic universe."

"You can calculate that the nugget that we believe perhaps gave rise to our universe was, maybe someone created it in their apartment in some other universe, was about roughly ten to the minus 26 centimeters across - 10 -26 cm., that's small, (Interviewer: "Really, really, small")..."

"You wouldn't really think that, intuitively, that you could build a whole universe from ten pounds of stuff, I think you would think that to build whole universe, I'm talking about a universe with stars and galaxies, hundreds of billions of star and hundreds of billions of galaxies, you'd think that you'd need more than ten pounds of this stuff, but it turns out that that's all you need, because the repulsive side of gravity is so powerful that it actually injects energy from gravity itself into the expanding space, so that, from that point of view, all you need is the seed, and then gravity takes over and does the rest of the work."

Further, Dr. Greene continues in this interview that the proper "seed" is a "black hole":

"You give me any object, and if I squeeze it sufficiently small, then according to the classic rules of General Relativity, if you make it small enough, it will be a tiny black hole. . . Seriously, there is nothing that you can give me that I couldn't turn into a black hole by squeezing it sufficiently small...There are processes, where particles can slam into each other, at very high energies, and the calculations show that if they slam to together at sufficiently high energies in the right geometric configuration, they can create a tiny black hole and this is not just hypothetical; there is a new machine, in Geneva Switzerland, called the Large Hadron Collider, and one of the things that may happen at the Large Hadron Collider is the creation of microscopic black holes in the collisions between protons and protons; these will be tiny black holes, but black holes nonetheless."

When asked, in the interview, "wouldn't that blow up everything?", Dr. Greene says:

"People have studied that issue in great detail, and found that, *at least according to the proposals that are now on the table* about how in principle you would create a universe, that wouldn't be a worry, that wouldn't happen." (*Italics* added)

So, we have the most popular physicist in the Western world stating that, yes, the LHC is being formulated to establish "black holes" (which eat up everything), but, no, *at least according to the proposals that are now on the table,* our world and every living thing in it will not, so this "current opinion" goes, be destroyed by this Large Hadron experiment. Thus, the fate of the world is being wagered on a matter of opinion.

The federal lawsuit, eventually dumped on technical grounds, brought out faint public echos of one of the Academy's taboo subjects; whether there are risks to humanity (and all animals, and also all the plants, all the insects) in the LHC project. Such rare sprinklings of warning are faint counterpoints to the praise commonly encountered when discussing this project. For example, the January 2008 issue of *Popular Science* offered *bookmaker's odds* on the potential findings from the impending Large Hadron Collider experiments.

These bookie odds included, at one end; A) a "9 in 10" bet that the Higgs Boson (a long sought sub-atomic particle) would be found, to; B) a "3 in 5" bet that Dark Matter would be located, to a purported; C) "1 in 10 million" bet that parallel universes would be found. Finally, there was Exacta wager, sub-titled "COSMIC IMPLOSION" in which the author blithely wagered the odds at "1 in a googol" that: "If the fundamental vacuum state of the universe is not perfectly stable, the LHC could make it decay, taking everything in the cosmos with it."[33] That mere possibility leads me to wonder whether, given trillions of stars, the encounter of black holes in binary systems might in some instances have been traceable to pride.

Chapter 12
About Surprise

Mankind's capacity for accurate scientific prediction of complex future events is limited. A few high profile examples easily illustrate this point: 1) The Gemini launchpad explosion, killing all on board; 2) the explosion of Columbia, killing all on board; 3) the loss of Challenger, killing all on board; 4) various premature declarations regarding Iraq, where so many have died; 5) the escaped tiger story in San Francisco, and, 6) let us not forget The Titanic. Some of the science-based bad calls would seem darkly comedic were it not for the costs involved. It is publicly stated by NASA that the $125 million dollar Mars Climate Orbiter was lost in 1999 because NASA's Jet Propulsion Laboratory and its spacecraft team at Lockheed Martin were using different measuring systems; Lockheed was using inches, while JPL was using the metric system. Despite this difference, the spacecraft was in orbit around Mars before the mistake was discovered, so NASA's story goes.

All of us experienced some level of world view alteration, for example, between September 10, 2001 and September 12, 2001. The unexpected occurrence, experience has taught, is often also the most profound.

The unpredictability of complex systems is not merely a factor learned the hard way by wannabe stock market speculators and new visitors to the race track; there is a developed science in this area, including but not limited to the area of scientific thought typically referred to as Chaos, as well as recently published works, such as Nassim Nicholas Taleb's writings on outlier effects[34], which writings question the reliability of Bell Curve predictability. These studied works indicate that the sureness many of us tend to feel about the reliability of scientific thought itself is often not justified, which cuts to risk analysis.

The phrase "outlier outcomes" refers to the results taking place which were outside the expectation paradigms of the people responsible for estimating possible bad outcomes, as in, "*I didn't know it was loaded.*" This is important in the collider context because there is no miner's canary available to give early warning of outlier results: Outlier events are outside of the expected, and may be sudden, or may be subtle and unnoticed, and yet ultimately of great consequence.

In order to discuss the effects of Chaos and Outlier risk assessment in complex systems, it will be useful to first visit the reasoning style which underlies our currently generally accepted scientific way of thinking, which is Gaussian analysis, most publicly thought of as Bell Curve analysis.

Most of us were at some point taught to visualize Newtonian physics in terms of a pool table: Under the deterministic laws of Newtonian motion, calculations could predict the exact outcome of any angular interface between bodies in motion, including, in the astronomical context, the interplay between gravitational relationships.

In the Newtonian era of perception there was a belief in the interplay of absolute factors. Norbert Wiener said that Newtonian physics "described a universe in which everything happened precisely according to law, a compact, tightly organized universe in which the whole future depends strictly upon the whole past."[35]

We have seen, though, that this is not the case. In a remarkable and multifaceted discussion of the limitations inherent in estimating future risk from past events, Peter Bernstein, in *Against The Gods, The Remarkable Story of Risk*, observes that:

> "Over time, the controversy between quantification based on observations of the past and subjective degrees of belief has taken on a deeper significance. The mathematically driven apparatus of modern risk management contains the seeds of a dehumanizing and self-destructive technology. Nobel laureate Kenneth Arrow has warned, '[O]ur knowledge of the way things work, in society or in nature, comes trailing clouds of vagueness. Vast ills have followed a belief in certainty.' In the process of breaking free from the past we may have become slaves of a new religion, a creed that is just as implacable, confining, and arbitrary as the old.
>
> Our lives teem with numbers, but we sometimes forget that numbers are only tools. They have no soul; they may indeed become fetishes. Many of our most critical decisions are made by computers, contraptions that devour numbers like voracious monsters and insist on being nourished with ever-greater quantities of digits to crunch, digest, and spew back."[36]

As is mentioned elsewhere, Planck and Einstein remained in disagreement for years, over whether the "Planck Constant" which Planck found an essential mathematical tool, did in fact, as Einstein contended in revolution, correspond exactly to the photon, a real object, thought of in Relativity as a particle, and viewed still as that, due to the capacity of gravity to bend its path.

Yet, sometimes we, as individual people, and much more so in groups, can get lost in a confusion between the symbol and that which it was originally articulated to represent. The slogan, thus, is not the candidate. Here, as to the LHC, there are incredibly elaborate numerical calculations, which, as Dr. Lee Smolin, supra, has noted, may be so far into the abstract as to have lost connection with corporeal practicality. Where risk of loss is concerned, particularly catastrophic loss, such a divergence can be, as was the case in the Mars Climate Orbiter, the harbinger of disaster.

Chapter 13
Physics Throws A Curve

Both in theories and in experimental verification, an outlook grounded purely in Newton has been difficult to sustain for roughly a hundred years, at the very least since Eddington's proof of Relativity in 1919. Some differing basic theories of physics after Einstein will be surveyed, but at a later point. The focus next is not on the scientific players or their plays, but rather on the playing field of consensus which underlies the cosmological dialogue, which is the Gaussian curvature mechanism of reasoning.

LaPlace's *Analytic Theory of Probability*, published in 1812 is often viewed as seminal to the acceptance of the "normal distribution" as a means of categorizing the likelihood of particular future outcomes[37]. The name that stuck, Gaussian, is after Carl Friedrich Gauss, who circa 1798, applied a concept of normal distribution to astronomical data[38]. Today, we use the name "Bell Curve" to describe this normal distribution. At what point, though, did the bell curvature become inherently intertwined with the predictive dialogue(s) of physics?

The physicist Willard Gibbs introduced the approach of employing a statistical likelihood of occurrence of events to the formulary way in which the interaction of those events, and resulting predictions of future events, were to be described. For immediate purposes, this brings us near to the "modern" scientific version of reality quantification. It is a way of quantification which we read about, and also one which we experience.

If both Gibbs and Wiener modernly get short shrift, perhaps it is in part because when Wiener wrote *The Human Use of Human Beings*, "chaos" had a very different meaning in science than it does today, essentially then as the consequence of entropy's defeat of system, and the background to that struggle. Today we see Chaos differently. Whatever the semantical reasons may or may not be, Wiener's work on Gibbs was the foundation of my own world outlook for decades, from 1970 forward until finally encountering Chaos, in the scientific sense of the word.

In the Newtonian approach, "time" was the stage, upon which all other factors acted out their parts. The interplay between objects in the universe, orbital geometries, for example, were just that, between objects, and the sole purpose of "time" was for the measurement of interval between interactions involving those objects.

Relativity changed all that, and the term and concept of "Spacetime" developed in the scientific and then the public consciousness. Even kids in high school now "get it" that, with Relativity, there is no division between the "clockwork" and the "stage." Time came to be understood as a function of events, not a backdrop to their existence. This is now broadly understood. Less common, though, is appreciation of the change to a probabilistic world view. On this Norbert Wiener wrote:

> "Many men have had intuitions well ahead of their time; and this is not least true in mathematical physics. Gibbs' introduction of probability into physics occurred well before there was an adequate theory of the sort of probability he needed. But for all these gaps it is, I am convinced, Gibbs rather than Einstein or Heisenberg or Planck to whom we must attribute the greatest revolution in twentieth century physics."[39]

What was that revolution?

If you've reached a certain age such that your conversations with doctors are more frequent than used to be the case, you have from time to time been told the statistical likelihoods, based on the general population experience, that you or a loved one will have a particular outcome from a particular procedure or medication. This was the revolutionary contribution of Gibbs' thinking, from what was "supposed" to happen (as in, it *will* happen" or "*will not* happen" in the Newtonian context) to a *probability quantified predictive process*, based upon the statistical experience of the larger group, applied to the hypothetical individual application and expected individual result. This has touched everything from ballistics to commercial bakery formulae. Gibbs' story survives in the work of Dr. Norbert Wiener, in particular, in his *The Human Use of Human Beings*. Dr. Wiener, amongst many other notable contributions, was the author of the term "cybernetics."

Dr. Wiener (1894-1964) received his Ph.D. from Harvard at the age of 19, and then studied at Cornell, Columbia, Cambridge (England), Gottingen, and Copenhagen. He was a Professor of Mathematics at MIT for forty years. In reference to a precursor to what we now generally think of as probability curvature thinking, in his *Human Use of Human Beings*, Wiener noted: "Borel, however, continued to maintain the importance of Lebesgue's work and his own as a physical tool; but I believe that I myself, in 1920, was the first person to apply the Lebesgue integral to a specific physical problem – that of Brownian Motion."[40]

Dr. Wiener's work was so central to my outlook in undergraduate and graduate education that for decades after that I relied entirely on the "probability curvature" outlook. When describing Gaussian predictability here, I mean "perceived overlapping curvatures of likelihood-based estimation synthesized to prediction of cumulative outcome" which is how I always looked at it before reading James Gleick. After James Gleick's *Chaos* in 1988, I appreciated new qualifiers to the Gaussian view.

The Gaussian outlook underlies the bookie odds in the *Popular Science* article; a world composed of percentages of outcome likelihood. Bell Curve analysis is either entirely reliable, or it is not. The possibility that Gaussian risk analysis is not absolutely reliable caused this essay, since, as the prose "cosmic implosion" illustrates, the LHC stakes are on the high side. It would be an easy cop-out at this point to sneer at *Popular Science* as a source, but in response to that hypothetical curled lip, it is noted that it is through such publications that the physics community sells those of us who foot the bill on how our tax dollars will be spent. On this general point, communications from the physics community to those of us who foot the bill, Dr. Lee Smolin noted:

> "In recent years many books and magazine articles for the general public have described the amazing new ideas that theoretical physicists have been working on. Some of these chronicles have been less than careful about explaining just how far the new ideas are from both experimental test and mathematical proof."[41]

The often-differing findings of deeply respected physicists will be discussed in the remainder of this book, but neither these essays as a body of work, or the sections which follow, presume to attempt answer of the many disputes between the Newtonian, Einsteinian, Quantum, and String, et al, views. Instead, it is the *fact* of these disputes which undercuts reliability of the calculus of risk, *because any dispute as to the ground rules for risk assessment necessarily results in limitation in the predictability of events stemming from those rules.* Two areas of well developed thought, Chaos and Outliers, mitigate against utter reliance upon statistical analysis as a predictor of future events in complex systems.

Chapter 14

A Brief History of Chaos

Edward N. Lorenz (1917-2008) was the first scientist to recognize and record Chaotic behavior in a system. As with hundreds of thousands of other science readers, I learned of this from James Gleick's *Chaos*. There is also a great graphic piece on this, called *Introducing Chaos*, by Sadar and Abrams, and for the diligent, there is Lorenz' own *The Essence of Chaos*.

A meteorologist, Lorenz was seeking a high degree of accuracy in weather prediction. In order to do so, he programmed a computer to emulate many elements contributive to weather systems. In the Gaussian analytical context, if sufficient data were present and programmed in, it was thought possible to develop highly accurate prediction of future events, since, so this outlook goes, when the same initial conditions were present, identical proximate results could be expected. Yet, today the weather remains full of surprises. The fact of these surprises is now understood from the vantage of Chaos Theory, which itself was helped along by legendary accident.

Legend has it that Lorenz went for a cup of coffee, having re-started his program in the middle, based upon previously programmed data. After his coffee, Lorenz returned to find that the computer had produced weather emulation results which were grandly inconsistent with the program's prior performance, yet on the basis of what appeared to be the same data. Lorenz then recognized that a very tiny numerical change had been made when the system was started at mid-program (.506 instead of .506127), and this tiny change produced a stunningly large difference in the subsequent predictions from the program's operation[42]. This is sometimes called, "unique susceptibility on initial conditions" or, more popularly, "The Butterfly Effect."

The first use of a crushed butterfly as the precursor to vast future changes pre-dated the Lorenz work and was in a 1952 science fiction story by Ray Bradbury in which a time traveling hunter, in panic, crushes a butterfly while in the hunt, only to return to the modern era to discover that a totalitarian personality has been elected President, a different person from the one elected prior to the departure from an ultra modern future to a dinosaurian past. Bradbury, as quoted in Dr. Ronald L. Mallet's remarkable physics autobiography *Time Traveler*, wrote that: "A small thing, that could upset balances and knock down a line of tiny dominos, and then big dominos, and then gigantic dominos, down all the years across time."[43] The Bradbury story is sometimes referenced by science writers in explaining Chaos, since it so colorfully illustrates the originally unexpected differences, out of anticipated scale, that very small occurrences can make in the evolving fate of those affected.

A major practical finding from Chaos study is that there are greater barriers to successful prediction of event outcome in complex systems than had been the conventional scientific outlook, prior to the 1963 publication of Lorenz' first article on this subject, "Deterministic Nonperiodic Flow" in the *Journal of the Atmospheric Sciences*[44]. Chaos Theory showed that the level of reliance previously placed in the predictive capacity of Bell Curve analysis was not justified.

There are many major disputes as to the nature of reality within physics today. Yet, as we have seen from the bookie odds, Gaussian prediction still provides the "playing field" upon which particle physics resides. Chaos Theory is teaching us, somewhat perhaps as the Gaussian view supplanted Newtonian determinism, that utter reliance upon curvature analysis for prediction is misplaced.

In addition to the impact of Chaos Theory, a further detractor from utter reliance comes from those who point to the potential risks from "outliers" namely risks which are so far outside the experimental paradigm that they were not considered, and yet which ultimately determined the outcome of the experimental process.

The fact that bookie odds have been dredged into a discussion of the LHC (not solely in *Popular Science*) compels mention of that most irksome of fellows, N. N. Taleb, whose works, *The Black Swan*, and *Fooled By Randomness*, in contrast to the *Popular Science* article, are heavily annotated. Taleb establishes that the Gaussian curvature method of estimation of future events is not only fatally flawed, but provably so.

Chapter Ten of *The Black Swan* is titled The Scandal of Prediction. Here, Taleb points not only to the inaccuracy of prediction as a regularly encountered, but psychologically denied, phenomena, but he also points to the hubris inherent in our culture, and perhaps in our species, which causes us to cling to a world of predictability which, as Dylan said in *To Ramona*, "just don't exist."[45]

Taleb mentions the building of the Sydney, Australia, Opera House, so unique in its sail-like design that you can probably visualize it now. The visibility part of the project was successful. However, the budgetary predictions for its construction were stunningly off. It was budgeted in 1963 to cost seven million dollars to build, but ended up ten years later costing one hundred and four million dollars, and yet for a lesser version of the originally planned building[46]. Such errors in prediction are not limited to politics, war or architecture, but rather have often been the case in hard science as well.

Ironically, some of the largest errors in prediction have been made regarding physics. For example, in 1987, the Reagan administration authorized, with Mr. Reagan cheering the physics community to "Throw Deep!"[47] the Superconducting Super Collider (SSC) project, calling for an 87 km ring and complex to be constructed at Waxahachie, Texas. The original cost estimate was $4.4 billion but, over time, the estimated cost of the machine rose to $8.25 billion, and by the Autumn of 1993, some estimates for construction were up to $11 billion dollars. Influenced in part by the science community, which feared the impact on other projects from such extreme funneling of resources into one project, Congress stopped the project, after spending $2 billion dollars on it, and digging a 22 km long hole[48]. Herman Wouk wrote a novel on it, called *A Hole in Texas* which is now on my reading list.

Taleb calls the arrogance of belief in prediction "epistemic arrogance" and his magician's hat is full of unpredicted rabbits, only a few of which need mention here. Taleb mentions the work of psychologist Philip Tetlock, which showed conclusively that experts' error rates were many times what they themselves had estimated, in fact, that there was no difference in the accuracy of predictive results whether one had a Ph.D., or an undergraduate degree.[49] Taleb points to the work of Makridakis and Hibbon, who tested the capacity of experts to accurately forecast real life developments. Their findings concluded that: "statistically complicated or complex methods do not necessarily provide more accurate forecasts than simple ones."[50] The example from Taleb's work which most hit home with me concerned an attempt by a governmental entity to predict the price of oil in 2004.

After a presentation, Taleb was approached by a governmental oil price forecaster who stated that, in January 2004, his department had predicted the price of oil for a twenty-five year period at an average of twenty-seven dollars per barrel. Six months later, around June 2004, they had to revise their estimate to fifty-four dollars per barrel, and it shot up from there[51]. The challenging part is that these same people, paid and expected to be forecasters, went back to forecasting oil prices twenty five years down stream, despite the abundant "real world" proof that such long range predictions simply were not possible due to error rate and factors which were not anticipated. In this regard see also Gleick's discussion of the turbulent regime[52]. To substantiate his point, Taleb footnotes reference to a 1970 forecast signed by the U. S. Secretaries of Treasury, State, Interior, and Defense, estimating that: "the standard price of foreign crude oil by 1980 may well decline and will in any event not experience a substantial increase."[53]   The facts show that probability based predictions in complex systems sometimes "get it wrong, very wrong."

Taleb relevantly reports that Poincare was suspicious of reliance upon the Gaussian approach to prediction, having written to one of his friends that; "physicists tended to use the Gaussian curve because they thought mathematicians believed it a mathematical necessity; mathematicians used it because they believed that physicists found it to be an empirical fact."[54]

As we will see in later chapters, that which physicists advertise as empirical fact will sometimes turn out to be neither.

Chapter 15

Applications

As to the LHC, Gaussian analysis was employed in the limited risk analysis so far conducted of the project and the experiment it houses; for example, the first CERN cosmology study was a probability calculation. As best articulated by Taleb in *The Black Swan,* such bell curve probability assessment is inherently subject to the potential for outlier outcome.

In fact, we definitively *know* that this is the case because (see Chapter 25) the most frequently referenced statistic, that there is a "1 in 50 million" chance of the destruction of the Earth by the LHC, as used by Sir Martin Rees at page 124 of his book, *Our Final Hour*, is a figure which was expressly derived from a paper by CERN theorists Arnon Dar, Alvarao de Rugula and Ulrich Heinz, which study uses the fact that the Earth and the cosmos have survived for several billion years to estimate the probability of colliders producing hypothetical particles called "strangelets" that might destroy our planet now[55]. If the risk analysis was Gaussian, as is certain, then the outlier possibility cannot be dismissed. Therefore, we should pause, and, without superstition, contemplate whether the LHC may demonstrate the greatest hubris to which our species has so far succumbed.

Fundamental issues in physics have sometimes been swept under the rug of public or academic consciousness by this Gaussian broom, as the cited mis-adventures show. Since the value of any "Bell Curve" analysis depends upon the reliability of the data going into the analytical process, our next excursion is into the realm of Quantum Physics. The question asked is: Do the physicists at least agree as to the basic rules of physics?

Chapter 16

There Is No Physics Consensus As To The Nature of Reality

Next discussed will be the mystery (a word chosen for technical accuracy) of the distinction between the mainstream Einsteinian analysis of physical questions, on one hand, and the also-mainstream quantum physics analysis on the other. While both of these outlooks are part of accepted, "mainstream" science, it is also certain, as admitted by leading physicists, that the Einsteinian and Quantum views are mutually exclusive.

The lack of consensus in physics which results from the controversy between these views of Nature is of independent analytical significance to the safety of the LHC experiment because the controversy by itself, whichever may or may not be "right" shows a schism in the reasoning of the physics establishment, and that is a limitation which strongly counts against their being allowed to play God with your fate, or mine.

Since at least the 1920's, it has been understood that Relativity, which works to explain existence above the molecular level, does not give a similar level of predictability at the smallest ranges of existence. Hundreds of volumes have been written around this general point. It appears broadly accepted in the physics community that, for example, an electron's path; 1) remains indeterminate until observed, and; 2) electron velocity and location cannot be simultaneously measured. A question thereby arises, as to where the electron is, if it is, when it is in its unobserved state. This question is one rootstock of the Many Worlds interpretation of quantum mechanics (and thus, necessarily existence) as postulated in 1957 by Everett[56], which itself has cautionary moral implications as to the risks of the LHC project. For those wishing to explore the concept of parallel universes, or, as called in physics, "the multiverse" Michio Kaku's book *Parallel Worlds* contains one of the clearest explanation of this hypothesis that I have encountered[57].

The high uncertainty in modern physics, as to the nature of the underlying universe(s) in which we live, is, by itself, a strong reason for restraint regarding any experiment which might endanger this universe and similar parallel universes.

Even from the outside, a voracious lay reader of the works of these physicists and their competing ideas and ideals can see clearly that there is a broad and deep chasm between Relativity and quantum mechanics. Yet despite an utter absence of consensus as to the nature of what is being done, a massively powerful experimentation, with uncertain result, is about to be undertaken. The current Gaussian means of risk analysis, given a divergence of views as to the very nature of the fabric of the universe, cannot be sufficient to adequately comprehend the possible extent of the risks involved.

Simply stated, it is not possible to assemble the puzzle of risk analysis if the pieces of that puzzle keep changing shape.

There are many scholars who provide essentially similar historical dialogues as to the evolution of thought after the experimental verification of Relativity. From my limited perspective, Roger Penrose, in *The Emperor's New Mind*, provides a particularly clear explanation of these steps, overlapping as they have been, in the evolution of thought in physics. Thus, Penrose states, in his discussion of "The 'Paradox' of Einstein, Podolsky and Rosen":

> "As has been mentioned at the beginning of this chapter, some of Albert Einstein's ideas were quite fundamental to the development of quantum theory. Recall that it was he who first put forward the concept of the 'photon' – the quantum of electromagnetic field – as early as 1905, out of which the idea of wave-particle duality was developed. (The concept of a 'boson' also was partly his, as were many other ideas, central to the theory.) Yet, Einstein could never accept that the theory which later developed from these ideas could be anything but provisional as a description of the physical world. His aversion to the probabilistic aspect of the theory is well known, and is encapsulated in his reply to one of Max Born's letters in 1926 (quoted in Pais 1982, p443):

> "Quantum mechanics is very impressive. But an inner voice tells me that is not yet the real thing. The theory produces a good deal but hardly brings us closer to the secret of the Old One. I am at all events convinced that *He* does not play dice.

> "However, it appears that, even more than this physical indeterminism, the thing which most troubled Einstein was an apparent *lack of objectivity* in the way that quantum theory seemed to have to be described. In my exposition of quantum theory I have taken pains to stress that the description of the world, as provided by the theory, is really quite an objective one, though often very strange and counter-intuitive. On the other hand, Bohr seems to have regarded the quantum state of a system (between measurements) as having no actual physical reality, acting merely as a summary of "one's knowledge" concerning that system. But might not different observers have different knowledge of a system, so the wave function would seem to be something essentially *subjective* – or 'all in the mind of the physicist'? Our marvelously precise physical picture of the world, as developed over many

centuries, must not be allowed to evaporate away completely; so Bohr needed to regard the world at the *classical* level as indeed having an objective reality. Yet there would be no 'reality' to the *quantum*-level states that seem to underlie it all."[58]

Those who note the imponderable nature of quantum theory are in very good company. Nobel laureate Richard Feynman said, "no one understands quantum mechanics" and; "its effects are impossible, absolutely impossible"[59] to explain based on human experience. Widely published and respected particle physicist Lee Smolin, whose comments regarding String Theory will be separately mentioned, commented regarding quantum physics that:

> "Quantum Theory is puzzling because it challenges our standard ideas about the relationship between theory and observer. The theory is indeed so puzzling that there is no universally accepted definition of it. There are many different points of view about what quantum theory really asserts about reality and its relationship to the observer. The founders of quantum theory, such as Einstein, Bohr, Heisenberg and Schrodinger, could not agree on these questions. There is now no more agreement about what quantum theory means than when Eienstein and Bohr first debated the question in the 1920's.[60]

The fact of this disagreement is illustrated today in writings on both sides of the String Theory debate. As Dr. Brian Greene stated at the start of his international best seller, *The Elegant Universe*: "As they are currently formulated, general Relativity and quantum mechanics *cannot both be right.*" (Italics in original)[61] As to String Theory (the generally subscribed outlook of most particle physicists supporting the LHC), Dr. Greene further stated in his *The Fabric of The Cosmos* that: "Even today, more than three decades after its initial articulation, most String Theory practitioners believe we still don't have a comprehensive answer to the rudimentary question, 'what is String Theory.'"[62]

If the leading physicists strongly disagree on such foundational points as whether Relativity or Quantum Mechanics is "right," then clearly no one can be certain at this time, based upon *evidentiary* reasoning, as to the true nature of the underlying fabric of the universe. This is variously dressed up, in which regard the Lorenz transform as to Relativity may be relevant[63], but the bottom line is that particle physics has resulted in deeply conflicting estimates as to the nature of existence itself: There is no objective scientific consensus as to the ultimate sub-atomic nature of existence: There are elaborate and diligently derived theories by deeply trained great minds, but this greatness has not as yet resulted in empirical definition, particularly as to String Theory.

There is a material difference between objective proof and informed conjecture, no matter how well informed: If a group of people who admittedly are without consensus as to the nature of empirical reality are engaged in making estimates as to the extent to which their experimental conduct may affect that reality; the potential for error rate mis-estimation is obvious.

Since the *Times* article about the Hawaii lawsuit, there has been expanding public buzz, in print and radio, over the issues which the plaintiffs sought to raise. The commentary from the physics community has aimed to calm alarmists, and to seek broad public recognition that the physicists really do know what they are doing, and for that reason, to pitch that adverse commentary about the LHC is by definition unscientific.

It appears to be the consensus of many in the physics community that the interior of black holes are "infinite." Terrence Witt, in his *Our Undiscovered Universe* surveys the several major schools of cosmological thought and then puts forth about 450 pages of formulae and calculations to support his own theory, Null Physics.[64] Without adopting Witt's position, which I have not as yet, his commentary on Big Bang cosmology illustratively cuts to the bone of any pose that the issue is permanently resolved:

> "Enter The Big Bang. Here the universe begins life as a submicroscopic particle of unimaginable density: a *universe particle*. This reduces the universe's size to a hypothetical (and by definition unobservable) realm even smaller than virtual particles. In so doing it connects it to the mysterious province of quantum reality where existence is nothing more than a statistical wave function. By its own admission, however, quantum physics has no underlying basis, so the virtual particle concept is a dead end, and the Big Bang focuses on the universe's birth as a problem of energy distribution, not energy conservation. But the size problem remains, unscathed by the paucity of the origin rationale. Either the universe came from somewhere, or it never came from anywhere. A sobering accounting problem is present in either case."[65]

The uncertainty admitted by all senior physicists as to the empirical proof of String Theory raises at least this question: Given the supposedly *infinite* nature of black holes, is it wise to chance the start up an infinite gravity sink in our three dimensional context which harbors life? Is it wise to create sub-miniature black holes if we already surmise that the universe actually started as a submicroscopic particle of unimaginable density? If even the potential exists that Terrence Witt is right and that, my paraphrase, the universe as we see it exists as interference patterns between the infinitely large and the infinitely small, would it not be wise to wait, at least until, if ever, Witt's Null Physics work is shown to be faulty math, before introducing into our context a subminiature exemplar of infinite smallness?

Witt's work may be no more fantastic than the hypothetical additional dimensions, which are a necessary rationale to String Theory. Is evaporation of sub-miniature black holes, if created, a sure thing, based on String Theory? Or is that evaporation a Gaussian probability, in which case, first, there is outlier risk, and, second: Under Many Worlds, the destruction of one or more universes within the multiverse would appear to necessarily result from the establishment of "mini" examples of "infinity" meaning in the null physics example, infinite smallness, here on Earth.

With these stakes not empirically verified: Is String Theory a reliable tool for risk analysis in the real world? This is a question in which we all have an interest.

Chapter 17

Strung Out

The LHC project will affect our physical reality, not just the realm of our ideas. There is nothing wrong with that, per se; physical actions cause physical changes in the world every day. It is hoped that the LHC, Atlas and all, will perform well, that no one will ever be hurt, that no environmental damage will ever result, and that something very worthwhile will be learned. In a world plagued by ignorance, starvation, war over resources, obsession, racism, and un-requited suffering from illness, the term "worthwhile" is subject to differing views; by its use here I mean "of substantial real value to the people of Earth."

As many scientific writings on the LHC detail, most of its collaborative supporters, and the theories underlying that which is expected from the experiment, are grounded in String Theory. To understand this politically correct vision as shared by most leading physicists, it is necessary to grasp the extent of its saturation of the physics community. Clearly, the fundamental belief system in particle physics today is String Theory. Dr. Lee Smolin wrote in *Quantum Gravity* that:

> "Measured sociologically, String Theory seems very healthy at the moment, with perhaps a thousand practitioners; loop quantum gravity is robust but much less populous, with about a hundred investigators; other directions, such as Penrose's twister theory, are still pursued by only a handful. But thirty years from now, all that will matter is which parts of which theory were right."[66]

In her book *Warped Passages*, Dr. Lisa Randall, the Harvard trained and vastly qualified physicist, discusses String Theory in order to describe the multi-dimensional view of reality to which she subscribes. No single paragraph or paragraphs from her monumental work should be singled out, let alone by a non-physicist like myself, as being Dr. Randall's sole view of the theory. Her book is full of wonderful puns, to help make her demanding subject understandable, and the section which Dr. Randall calls String Training contains the most lucid description of String Theory that I have so far encountered:

> "String Theory's view of the fundamental nature of matter differs significantly from that of traditional particle physics. According to String Theory, the most basic indivisible objects underlying all matter are *strings* – vibrating, one-dimensional loops or segments of energy. These strings, unlike violin strings, say, are not made up of atoms which are in turn made up of electrons and

nucleons which are in turn made up of quarks. In fact, the opposite is true. These are fundamental strings, which means that everything, including electrons and quarks, consists of their oscillations. According to String Theory, the yarn a cat plays with is made of atoms which are ultimately composed of the vibrations of strings.

"String Theory's radical hypothesis is that particles arise from the resonant oscillation mode of strings. Each and every particle corresponds to the vibrations of an underlying string, and the character of those vibrations determines the particle's properties. Because of the many ways in which strings can vibrate, a single string can give rise to many types of particles. Theorists initially thought that there was only a single type of fundamental string that is responsible for all known particles. But that picture has changed in the past few years, and now we believe that String Theory can contain different, independent, types of strings, each of which can oscillate in many possible ways."[67]

Dr. Randall is amongst those happily anticipating the possible benefits which may be forthcoming from the LHC. She makes twenty references to the LHC in her book to present the benefits she and other outstanding particle physicists hope to obtain through the LHC. For example, in discussion of the long sought Higgs Boson, Dr. Randall writes:

"The Higgs mechanism involves a field that physicists call the *Higgs field*. As we have seen, the fields of quantum field theory are objects that can produce particles anywhere in space. Each type of field generates its own particular type of particle. An electron field is the source of electrons, for example. Similarly, a Higgs field is the source of Higgs particles.

"As with heavy quarks and leptons, Higgs particles are so heavy that they aren't found in ordinary matter. But unlike heavy quarks and leptons, no one has ever observed the Higgs particles that the Higgs field would produce, even in experiments performed at high-energy accelerators. This doesn't mean that Higgs particles don't exist, just that Higgs particles are too heavy to have been produced with the energies that experiments have explored so far. Physicists expect that if Higgs particles exist, we'll create them in only a few years' time, when the higher-energy LHC collider comes into operation."[68]

Peter Woit graduated from Harvard in 1979, holds a Ph.D. in theoretical physics from Princeton University, and is a lecturer in mathematics at Columbia; he is part of the same faculty as Brian Greene. However, their views on String Theory differ greatly, as is discussed in Woit's book *Not Even Wrong*[69]. An endorsing quotation from Lawrence Krauss on its cover reads:

"Peter Woit presents an authoritative, sobering, and very readable history of a scientific and sociological phenomena that is largely unprecedented in the history of science. From a physics perspective, what is known as "String Theory" remains primarily the "hope" for a theory. Nevertheless it has dominated theoretical particle physics as well as the popular consciousness as few other notions have, all the while without ever making a single falsifiable prediction about nature!"[70]

Dr. Peter Woit mentions his own concerns, and the concerns of others, as to whether, given its combination of zealot belief and absence of empirical proof, String Theory is taking on the characteristics of a faith-based outlook :

"The qualms that many scientists have about Superstring Theory are often expressed as the worry that the theory may be in danger of becoming a religion rather than a science. Glashow is one physicist who has expressed such views publicly:

'Perhaps I have overstated the case made by string theorists in defence of their new version of medieval theology where angels are replaced by Calabi-Yau manifolds. The threat, however, is clear. For the first time ever, it is possible to see how our noble search could come to an end, and how Faith could replace Science once more.

"I have heard another version of this worry expressed by several physicists, that String Theory is becoming a cult, with Witten as its guru.....For instance, a string theorist on the faculty at Harvard used to end all his email with the line, 'Superstring/M-Theory is the language in which God wrote the world.'[71]

Public dialogue often centers on a dialectic often phrased as "science versus religion." In this dialogue, popular authors claiming to represent the scientific position (Dawkins, Harris, Hitchens, et al) will typically style their positions as empirical, and the positions of those defined as their opponents as superstitious. With String Theory we have a very different developing trend, towards a new phase, not of "science versus religion" but "science *as* religion." There are many aspects of organized group thinking, towards the promotion of group identity, which are common to all belief systems grounded in zeal, as opposed to empirically testable reality. Eric Hoffer's classic *The True Believer* concerns processes through which managed group thought is used to overwhelm dissent, such as by shunning, for example, Hoffer writes:

"The effacement of individual separateness must be thorough. In every act, however trivial, the individual must by some ritual associate himself with the congregation, the tribe, the party, etcetera. His joys and sorrows, his pride and confidence must spring from the fortunes and capacities of the group rather than from his individual prospects and abilities. Above all, he must never feel alone. Though stranded on a desert island, he must still feel that he is under the eyes of the group. To be cast out from the group should be the equivalent to being cut off from life."[72]

It is clear and broadly conceded in the physics community that there has been no empirical showing that the "strings" in String Theory actually exist. String Theory is, as admitted by its leading practitioners, utterly theoretical. Therefore, the objective factual situation here includes that: 1) There is a belief system, which is; 2) not empirically proven, yet; 3) which is receiving vast public and private support, and; 4) is being promoted, at the expense of positions based in experimentally verifiable science. In *Warped Passages,* Dr. Lisa Randall discusses the controversy, in the context of String Theory, between empirical science and purely theoretical science:

> "The conflict between the two scientific approaches is interesting because it reflects two very different ways of doing science. This division is the latest incarnation of a long debate in science. Do you follow the Platonic approach, which tries to gain insights from more fundamental truths, or take the Aristotelian approach, rooted in empirical observations? Do you take the top-down or bottom-up route?"[73]

The above language, "Do you take the top-down or bottom-up route?" reasons that the "top-down" approach is worthy of being followed in the pursuit of truth, to the exclusion of the experimental, because in the top down approach the search commences from "fundamental truths." If the same sort of statement were being made by a religious figure, many of us would regard it as "faith-based reasoning", historically as known as "divine right."

With all due respect, faith-based reasoning as a cohesive social force has often caused outlier results. In the LHC experiment we are entitled from the facts to trust the good faith of the vast super-majority of scientists engaged in the endeavor. Yet, when viewing any idealized construct which urges social action (such as vast expenditure) on the basis of ardently subscribed but non-empirical reasoning, we would be well advised to recall historian Paul Johnson's remarks on zealotry in his *Modern Times*, a *New York Times Best* Book of The Year, including that: "But the experience of the twentieth century shows emphatically that Utopianism is never far from gangsterism."[74] This is not stated to imply ill will as to any of these conscientious scientists but rather simply to note that reliance on criteria other than facts in the articulation of public policy has not historically been associated with ultimately effective strategic decisions.

In contrast to the untempered zeal of the ardent string theorists, Dr. Peter Woit is amongst those in the physics community who continue to insist on empirical proof. With his deep experimental credentials, Dr. Woit differs from those of his physics colleagues whose work has been utterly theoretical. In describing the failure of String Theory to obtain objective quantification, Dr. Woit explains pioneering scientific standards work of Karl Popper:

"The best-known proposed criterion for deciding what is science and what isn't is the criterion of falsifiability generally attributed to the philosopher Karl Popper. By this criterion an explanation is scientific if it can be used to make predictions of a kind that can be falsified, i.e., can be shown to be wrong. The falsifiability criterion can in some circumstances be slippery, because it may not always be clear what counts as a falsification. Observations may be theory-laden, since some sort of theory is needed even to describe what an experiment is seeing, but this problem doesn't seem to be at issue in this context....

"The standard model is an excellent example of a falsifiable theory, since it is one of the simplest possible models of its kind, and it can be used to generate an infinite set of predictions about the results of particle physics experiments, all of which in principle can be checked in an unambiguous way. On the other hand, Superstring Theory is at the moment unarguably an example of a theory that can't be falsified, since it makes no predictions. No one has come up with a model within the Superstring Theory framework that agrees with the known facts about particle physics. All attempts to do so have led to very complicated constructions that show every sign of being the sort of thing one gets when one tries to make an inappropriate theoretical framework fit experimental results."[75]

Does it matter that the risks inherent in the operation of the LHC are being assessed by String Theorists? String Theorists, who have encouraged the development of the LHC and now run it, anticipate the possible occurrence - and then hoped for evaporation - of tiny black holes as a result of this summer's experiment. *It is respectfully suggested that the lack of empirical proof for String Theory be taken into account in assessing the level of reliance which to be placed in the risk assessment process now in place at CERN.*

We are about to send protons hurling towards one another very close to the speed of light, in order to break them into component pieces, admitting while doing so that we don't know what those pieces will be, and hypothesizing that black holes with "no more energy than a mosquito" may result. The String Theory outlook provides a framework for many solutions. However, it is irrefutably clear that it is an unproven theory. This gives rise to questions along the line of, bluntly: Should we proceed to de-construct matter under the direction of people who demonstrably do not know what they are doing?

We have no data indicating that the LHC is being constructed on the basis of any underlying theoretical framework which has not yet migrated into public view; all public data indicate that the string theorists are at the helm of the project, which reduces down to a collection of facts indicating that we have a huge and expensive program in the hands of people who do not have an empirical basis to attest to either its efficacy or its safety. If these ardently held beliefs, with a similar absence of hard factual proof, were stated from a pulpit instead of a university, many of us would view the words spoken as tinged with obsession.

A critical test of the validity of an experimental finding is whether other researchers, following the same experimental protocols, can routinely obtain the same finding. If so, the finding is said to be "replicable" which indicates that the result of the experiment, for one example, was not the result of procedural error in an element of an experiment, or systemic error in the experiment as a whole. As discussed by Dr. Dean Radin in *The Conscious Universe*, using meta-analysis techniques..."psychologist Larry Hedges of the University of Chicago discovered a surprising result; some experiments in the soft sciences are as replicable as those in the hard sciences..."[76] which research by Hedges supported the conclusion that: "even in particle physics, one of the most rigorous, well-funded, and hardest of the hard sciences, the *actual* replication rates are comparable to the replication rates observed in the soft, pliable work of the behavioral sciences."[77]

In particular, Dr. Radin reports that the Particle Data Group (PDG) of the American Physical Society, in providing reviews of experimental findings, uses subjective, and in any event reviewer-defined, means for picking and choosing what elements of data will actually be reported. Thus: "Among the reasons listed for discarding data are; 'The results involve some assumptions we do not wish to incorporate,' and 'The measurement is clearly inconsistent with other results which appear to be highly reliable.' In other words, data are discarded to reduce 'outliers' that are thought to be flawed in some way. As new data are added to the old, the claimed precision of the claimed estimate increases. However, as the PDG writes: 'Some cases of wild fluctuation are shown; this usually represents the introduction of significant new data or the discarding of some older data. Older data are sometimes discarded in favor of more modern data if it is felt that the newer data has fewer systematic errors...By and large, a full scan of our history plots shows a rather dull progression towards greater precision at a central value completely consistent with the first data point shown.'"[78]. But, as Dr. Radin then points out: "Rather dull, that is, *only if* the outliers are removed."[79]

There are related bottom line issues here. I started reading physics many years ago with the starting assumption, which seems to be common, that physics was a "hard" science, in which true factual objectivity would dominate, and subjective judgments would not have much of a voice. That is what I liked about it. It is clear from the very core of particle physics reportage, the PDG, that this is not true, that despite the lab-coat vernacular used by practitioners and fans to describe the process, the reality is that subjective judgments are an integrated part of physics reportage at the highest levels in the field. Secondly, as shown from inquiry as to String Theory, there is no objective, or to use Popper's term, falsifiable, proof that String Theory has any basis in fact. Thirdly, there is very substantial disagreement between the elite minds of the physics community (here by "elite" I do not mean "most popular" but rather refer to the work of credible Ph.D.'s who have devoted their lives to the field and yet harbor very different beliefs about the nature of reality). Fourthly, it is abundantly clear that in physics, just as in all other communities, there is a tendency to brand thought which is popular within the community as acceptable, while branding less popular ideas as not acceptable, even though in some instances, such as the Experimental view versus the String Theory view, it is those of the less popular position which actually have what those of us practicing trial law would refer to as "evidence."

The diligent reader will encounter conjecture and dispute amongst particle physicists as to the true nature of existence. The certain fact in particle physics is that its practitioners disagree with one another, often, as Dr. Smolin has indicated, with vehemence. Further, at least one top publication in particle physics, which *defines the acceptable consensus reality of physics thought*, shows very clearly, as shown by Radin's quotation of Hedges' work, that subjective factors result in the filtering of outlier data out of the baseline paradigms. On the basis of these concerns, the risk assessments attached by physicists to their work are necessarily uncertain. Normally, this doesn't matter much, what's a building here or there, after all. However, with the LHC, there is vastly more at stake. The question which this naturally brings up is whether there is reason for concern, in regard to the Large Hadron Collider experiment, that we have folks tinkering with the very underlying matrix of our existence itself, who admit, as they tinker with our real world, that their positions are purely theoretical, and in some instances mutually exclusive.

The issues discussed here go beyond the question of whether an adverse result will come from the LHC. These issues include the ethical consideration as to whether the world of non-physicists should have a voice in physics experiments which might affect them; a contrary position where uncertainty resides would be difficult to morally defend.

Yet, physics experimentation has in the past brought us utterly unexpected understandings of our universe, which, in ways both practical and not, have helped us to understand our surroundings, and in some instances avoid superstitious ideation, a good thing. Sometimes, such as regarding entanglement, the physics experimentation has disclosed proof of ideas in quantum mechanics which seemed at the time of their announcement to be beyond the pale of reason in the context of locality. For reasons to be discussed, entanglement is arguably an issue relevant to risk analysis here, so a brief survey follows.

Chapter 18

An Entangled Web

I was privileged recently to listen to an astrophysicist, a leader in his area of study. Amongst the exhibits were astronomical images and data which were convincing to him, and thus easily good enough for me, that the bending of light had been authenticated in the astronomical context in extreme gravity, just one more proof, of the many since Eddington[80], that Einstein "got it right" in the great majority of his towering work.

Yet the history of Relativity shows that there was more to the universe than Dr. Einstein believed. The most dramatic example of this is non-locality, resulting from entanglement, which is counter-intuitive, both from the standpoint of the every day reasoning that we use to operate in the world, and in the context of Relativity beliefs. It is a rent in the original fabric of Relativity. The concept can be summarized in a few paragraphs, and that is all that will be attempted here.

Black-letter Relativity theory demands that light speed be an absolute, meaning that nothing can go faster. Pages, books even, could be spent defending that point, but it is very well accepted, including by string theorists.

In the ordinary position of an audience member with a question, I once asked Dr. Greene, at Dominican University, whether in light of later slit interferometer experiments seeking to replicate the 1887 experiments of Michelson and Morely[81] there might be any legitimate challenge to the Michelson-Morely findings, which Einstein acknowledged as vital to Relativity. Michelson-Morely was the experiment which determined the speed of light, and ruled out the previous widely held concept of the "aether" as the lowest common denominator of existence. I was aware that it is axiomatic in modern mainstream physics that Dayton Miller's careful work, as published in *Reviews of Modern Physics* in 1933, suffered from a systemic error having to do with the temperature of the light measured, as confirmed by the work of Shankland, McCuskey, Leone, and Kuerti in 1955[82], but I was nonetheless interested in Dr. Greene's comment on the issue.

Dr. Greene quickly dismissed the question by explaining that Michelson-Morely had been replicated many times, in regard to light speed. However, when I briefly clarified, "No, I was speaking as to entanglement," Dr. Greene then said, "that will be the last question" and launched into an incredibly poetic dialogue on entanglement, which, briefly, shows a great interconnectedness between all of us. I will not attempt to discuss any proposed relationship between Miller's work and entanglement here.

I could not begin to do the subject of entanglement justice here, but as a starting point, quantum mechanics, as to the formation of matter, appeared to Dr. Einstein to require an impossible result, which result violated both Relativity and common sense, namely that: <u>Atoms, once connected, would remain sympathetic to one another forever; the word sympathetic is not used here as emotive.</u>

One way of saying that our universe makes sense is to note that there is never a result without a cause, and that, in Relativity, all causes act at, or below, the speed of light. If it were possible for a cause and its effect to be simultaneous, over a long distance, this would violate Relativity, since some moment of time, however infinitesimal, will always be necessary for the communication of the cause to the effect, and time is limited in Relativity, by light. Think of a bullet in motion, for example, or the time delay experienced by lunar astronauts in their conversations with mission control; the bullet never arrives the exact instant it is shot, and radio signals, traveling at the speed of light, always take some element of time to get from the place of origin to the place of receipt. This concept is central to many of our great technological innovations, such as the older (and beloved to me) Loran-C navigation system, and the GPS system which has now largely supplanted it.

In 1935, Einstein, Podolsky, and Rosen published a paper which postulated that, if quantum mechanics really worked as Niels Bohr claimed, this would require non-local action, the simultaneity of cause and effect, over a distance, in utter violation of Relativity.[83] Such a possibility was derisively called by Einstein as "spukhafte Fernwirkung" or "spooky action at a distance."[84] The Einstein-Podolsky-Rosen paper said that non-locality was not possible.

But it was. As described by Dr. Dean Radin: In 1964 Irish physicist John Bell mathematically proved that quantum theory requires "spooky action at a distance" and that non-local action really does take place. The famous proof would become known as Bell's Theorem[85], and some physicists would regard it as the most profound scientific discovery of the twentieth century. The popular work of Lynne McTaggart, *The Field*[86], is the best door I've found into the rabbit hole of non-locality, but regardless of which of the huge number of articles you might personally prefer, suffice for present purposes to state that non-locality has now been experimentally verified, and mathematically proven and since it actually works, in the contest between Relativity and quantum mechanics on the point, the quantum mechanics view was experimentally verified.

Rosenblum and Kuttner, in their discussion of non-locality explain that entanglement has been proven in the laboratory setting, first by John Clauser, as a post-doctoral graduate student of University of California Berkeley, in the 1970's, and then more famously by Alain Aspect near Paris in the 1980's. Entanglement has been empirically established as a real phenomenon, as predicted through quantum mechanics.

In addition to its mathematical proof, non-locality is now viewed by many competent experimenters as a likely road towards explanation of phenomena, such as pre-cognition, which have now been copiously experimentally verified, but which findings are sometimes rejected on the basis of reasoning to the effect that it "just can't be." Hundreds of thousands of very carefully designed and operated experiments, as discussed by Dr. Dean Radin in his *The Conscious Universe*, and also his *Entangled Minds,* including the pioneering work at Princeton of Jahn and Dunn[87], in the Princeton Engineering Anomalies Research (PEAR) lab, and the work of Drs. Hal Puthoff and Russell Targ[88] at the Stanford Research Institute, have experimentally and/or statistically demonstrated beyond rational dispute that human minds from time to time perceive events or thoughts by non-local means. It is an interesting anomaly in physics today that String Theory, for which there exists no hard evidence, is more generally viewed as "scientific", yet the painstaking work of experimenters such as Jahn, Dunn, Puthoff, Targ, and the late physicist David Bohm, in his *Wholeness and the Implicate Order*, which has postulated a holographic understanding of the universe, are less "mainstream" even though well supported by substantial valid experimental findings[89]. However, while consciousness experimentation is one area tending to show the existence of non-locality (which I believe will someday be understood as key to gravity), one need not rely solely on that area of experimentation, since the work of Bell is a reliable academic proof. If one is to be genuinely scientific, it must be accepted based upon Bell's findings, that there is a great underlying connectedness in our universe. It was not until listening to Brian Green's truly elegant lecture on entanglement, that night at Dominican College, that I comprehended that universal entanglement, as a necessary result of the conclusion that all matter, as we know it, was in a super compressed precursor state, prior to the Big Bang.

Chapter 19

The Barbershop Mirrors

Many of the scientists who postulate a further existence underlying the four dimensional space, which we visualize and live in, are string theorists; but the view of a hidden underlying order is not by any means limited to the String Theory vision. For example, the work of physicist David Bohm, as summarized for the interested lay reader in his *Wholeness and the Implicate Order*, may support alternatives to the current, ardently held, if young, belief that the underlying nature of the universe is truly settled, and that String Theory is the answer. Bohm commented on the existence of orderly symphony in the movement of electrons, for example, yet without any three dimensional explanation for the level of coordinated movement observed. Bohm's work elegantly enquires as to whether or not there are hidden factors affecting our relativistic universe, and he answers that question in the affirmative, postulating, along with others, that a "holographic" model of existence which we currently perceive is, very roughly stated here, complex interference pattern projection from an underlying universe. There is an excellent popular discussion of the holographic viewpoint in physics in Michael Talbot's *The Holographic Universe*.

Considerable legitimate scientific inquiry has lead to the conclusion of multiple dimensions, unseen by mankind, but necessarily present in the mathematically derived investigations of top physicists, such as by Lisa Randall, Ph.D., in her *Warped Passages*.

Dr. Randall, a leading string theorist, postulates that the underlying nature of reality, based on String Theory calculations, necessarily involves at least eleven other dimensions, beyond those of height, width, depth and time that we perceive in the four dimensional universe of our senses. Amongst Dr. Randall's views is the possibility that "we could be living in an isolated pocket of space that appears to be four dimensional."[90]

An unanswered question: Might our "isolated pocket" be fragile?

In fact, even before the very young (thirty year) era of String Theory, physicists told us that we were possibly in a multi-dimensional barbershop mirror of reality, with adjacent parallel dimensions existing simultaneously with that which we perceive. This is the "many worlds" or "parallel worlds" hypothesis, as articulated as a logical extension of quantum physics by Hugh Everett III. Physicist Fred Alan Wolf states the "many worlds" hypothesis as follows:

"The 'many worlds,' or 'parallel universes' version of quantum physics states that the observer, in observing, is actually becoming part of the observed by noticing and remembering what he or she experiences. If a quantum system is capable of being observed in one of several possible states, then when an observation occurs, the system enters all of these states and the observer's mind splits into a companion state associated with each possible physical state of the system."[91]

Chapter 20

A Quantum Of Weirdness

Richard Feynman has remained a hero to me for more than a quarter century, ever since my reading of his sharp and funny autobiographical book, *Surely You're Joking Mr. Feynman* in the late 1970's. *Surely You're Joking* inspired me to read more about Dr. Feynman, including, with humility, my attempt at ingestion of his 1964 Lecture on the motion of the planets, heroically preserved and, as to the blackboard demonstrations, re-constructed in *Feynman's Lost Lecture*, by David L. Goodstein and Judith R. Goodstein, and available in a wonderful package telling the story of the incredibly dedicated work of these authors in preserving this precious gem and burnishing this precious gem.

I remained a Feynman fan, both during his life and since his death in 1988, one unimportant admirer amongst at least hundreds of thousands of people who have found his towering intellect, personal forthrightness and humor admirable.

Dr. Feynman had a gift of getting across large complex ideas in a concise way, such as the his previously cited comments on Quantum Physics: "no one understands quantum mechanics" and "its effects are impossible, absolutely impossible"[92]. Yet, even if Quantum Physics is "impossible" still, it is important for we lay readers to appreciate what these highly specialized physicists are talking about. Here, again, Dr. Feynman comes to our rescue, with his remark that the double-slit experiment captures the entire problem of quantum mechanics[93].

The double slit experiment is the tap root of quantum physics because it is the experimental result which appears to give empirical proof to the "dual" theory of subatomic particles, namely that sometimes particles, such as photons or protons, act as we'd expect from particles, and other times, it appears that they act as we would *only* expect of waves. This in turn has given rise to a whole host of physics holy books promoting quantum physics, which can only be challenged at the risk of ridicule. To get a handle on these materials, we need to take a look at the double slit experiment.

The double-slit experiment is the most widely described experiment relating to quantum mechanics. Most books referencing quantum physics which were written for the intelligent lay reader. For our purposes, the double-slit can be visualized using everyday images.

The double slit experiment is carried out with a particle broadcast source, a solid plate with two slits in it, and a screen. The slitted plate is placed between the particle source and the screen. Okay, now let's darken the room, and shine a light through the slits. We know what to expect, right? There are two holes, so we expect that there will be two areas of illumination on the screen behind the plate, right? Logic and our ordinary intuition tell us "a light, two holes, therefore two illuminated areas."

It is important to this discussion to keep in mind that light has, since a period of experiments conducted between early 1919 and late 1923, and verified many times after, been recognized as being composed of *particles*, called "photons." A particularly good discussion of the early experimental confirmation of the particle nature of light can be found in Paul Johnson's epic work, *Modern Times*, describing first the famed outcome of Eddington's photographs of an eclipse of the sun by the moon, at the Canary Islands:

> "The expedition satisfied two of Einstein's tests, which were reconfirmed by W. W. Campbell during the September 1922 eclipse. It was a measure of Einstein's scientific rigor that he refused to accept that his own theory was valid until the third test (the 'red shift') was met. 'If it were proved that this effect does not exist in nature,' he wrote to Eddington on 15 December 1919, 'then the whole theory would have to be abandoned.' In fact the 'red shift' was confirmed by the Mount Wilson observatory in 1923, and thereafter empirical proof of Relativity theory accumulated steadily."

For our present purposes, it is sufficient to note that rigorous testing has established that light is directly affected in its path by the gravitational effect of massive bodies, and thus it is further fair to state as accepted that it, in that context, acts as a *particle*. Therefore, returning to our double slit experiment, when we send a stream of particles through two adjacent slits in a plate, onto a screen behind, we would expect a resulting photograph taken by film on that screen to show the presence of two illuminated areas.

However, when a particle source, including a proton source, is used, four (or more) areas of illumination are shown on the screen. Why?

The problem created for physicists is that, while two areas of illumination signature would be expected for *particles*, a larger number, four or more, is routinely expected only when *waves* are involved, due to the effect of "interference patterns" between those waves. Unfortunately for both simple logic and the intuitive expectation, the "interference pattern" result, associated with waves, *also occurs with photons*. Briefly treated here, and widely available elsewhere, the bottom line is that this anomaly has led to the "particle/wave duality" ascribed to light, in the most simple terms, that it sometimes acts as a particle, and sometimes as a wave. This is the rootstock of Quantum Physics.

One of Albert Einstein's great accomplishments was to rigorously define the photon as a tiny particle of light, bearing a size exactly equal to what had before him been sometimes known as the "Planck Constant." Walter Isaacson, in his intimate 2007 biography *Einstein, His Life And His Universe,* describes the scientific papers and correspondence through which Einstein appeared to have settled upon a "particle" theory of light, and a brief portion of this monumental work describes the controversy, which, as we shall see, persists today:

> "In effect, as Einstein noted in a paper the following year, his role was that he figured out the physical significance of what Planck had discovered. For Planck, a reluctant revolutionary, the quantum was a mathematical contrivance that explained how energy was emitted and absorbed when it interacted with matter. But he did not see that it related to a physical reality that was inherent in the nature of light, and in the electromagnetic field itself. 'One can interpret Planck's 1900 paper to mean only that the quantum hypothesis is used as a *mathematical* contrivance introduced in order to calculate a statistical distribution, not as a new *physical* assumpton,' write science historians Gerald Horton and Steven Brush.
>
> Einstein, on the other hand, considered the light quantum to be a feature of reality, a perplexing, pesky, mysterious, and sometimes maddening quirk in the cosmos. For him, these quanta of energy (which in 1926 were named photons) existed even when light was moving through a vacuum. 'We wish to show that Mr. Planck's determination of the elementary quanta is to some extent independent of his theory of blackbody radiation" he wrote. In other words, Einstein argued that the particulate nature of light was a property of the light itself and not just some description of how light interacts with matter.
>
> Even after Einstein published his paper, Planck did not accept his leap. Two years later, Planck warned the young patent clerk that he had gone too far, and that quanta described a process that occurred during emission or absorption, rather than some real property of radiation in a vacuum. 'I do not seek the meaning of the 'quantum of action' (light quantum) in the vacuum but at the site of absorption and emission, he advised.'" (Citations omitted)[94]

The discussion of the particle versus wave character of the photon has remained with us for these last mere hundred years. When shooting light through two slits onto a photographic plate, common sense tells us that we'll obviously see changes in the developed film showing that two areas were illuminated on the target screen when the light source was switched on, but in reality four light exposed areas are found. The problem is, that is what we'd expect with *waves*, *but not particles.*

Waves behave that way because of their reflections have a sequestering effect on the original signal which results from *interference patterns.* So, that's the problem: The double slit uses what Einstein has told us are particles. In real life, light acts in a manner consistent with particles, such as in travel through a vacuum, and yet, on the other hand, the double-slit shows us that what we have for 83 years (as of 2009) called "photons" sometimes act like particles, and sometimes acting like waves. This dichotomy is at the heart of quantum mechanics, and is one of the major sources of "quantum weirdness" which is a scientific label for a big collection of anomalies that no person in the world of public physics yet fully understands.

Eighty-three years is not all that long a period of time, and it is suggested that we should not believe in something with religious zeal, as though empirically established for centuries, when it is both recent and only partly understood.

In contemplating the LHC, the mere fact of the gaps in scientific knowledge is significant, as it cuts against the propriety of arrogance by physicists.

We, the lay readers, are morally and logically entitled to regard a physicist's writing for a general audience to be a truthful representation of that physicist's work. For example, if Michio Kaku, a very highly regarded physicist, presents arguments in support of the existence of multiple adjacent dimensions, we are all entitled to treat those arguments as Michio Kaku's truth on the subject, so long as is he, or any other author, is accurately quoted.

We lay readers, who by our tax dollars, donations, and political endorsement, give support to the scientific community, are entitled to take the scientists at their respective words, the conclusions, stated in English, which represent the end products of their computations, observations, and research. Scientists have an obligation to be straight with us, for many reasons, including not only our safety, but that we pay the bills.

Thus, if there are (as is the case) wide divergences in the published opinions of scientists regarding fundamental points in physics, we, the general readers, and entitled to fairly quote both sides of those divergences, and to note mutual exclusivities when they are found. This general point is important to any book reporting on physics, including this one, but, as will be seen, it becomes very important when we compare, for example, the quantum physics viewpoint which has grown out of the double slit experiment, next described, with the work of Dr. Lewis Little. The following is an excerpt of a reader's review of Dr. Little's work, which was posted by Joshua J. Hansen on Amazon as of Sept 2009:

"For a century, experiments with subatomic particles – such as the double-slit experiment linked to the particle-wave theory of light – have yielded puzzling results. Physicists have long possessed equations that successfully predict these results, but quantum mechanics, the theory that has accompanied those equations, is less satisfying. Like the math, it does explain the results, but only if the physicist accepts such propositions as these:

- A single particle at once occupies multiple, separate locations.
- A particle travels on multiple paths toward multiple destinations but arrives at only one destination.
- The destination of a particle depends on its conscious observer. (Consciousness determines reality.)
- Conscious observation of a particle collapses it from a "superposition of multiple states" into a single state.
- At any given instant, a particle can have only an exact position or an exact velocity, not both.
- Objects can interact from a distance without any physical means to do so.
- Objects can change location instantly (moving at infinite speed).
- Causation can run backward in time (so that an event in the present determines what happened in the past).

Unlike Dr. Little – and unlike Albert Einstein before him – most physicists accepted such propositions. They had no choice: that no other theory had been proposed proved that quantum mechanics was the correct theory, and the illogical nature of this theory proved that the rules of logic were delusions. Many of the physicists who reached this conclusion no doubt had I.Q.s over 200, like Dr. Little's. These physicists did not, however, match Dr. Little in intellectual discipline or persistence.

As a disciplined thinker, Dr. Little knew that quantum mechanics had to be wrong. As a persistent investigator, he continued his inquiry until he discovered a likely source of the error. Earlier physicists had labored under an unrecognized assumption: that the wave on which a particle of light travels runs in the same direction as the particle. Dr. Little recognized this assumption as an assumption and tried working with the opposite notion: that the wave runs from the particle's destination to the particle's source. This alternative notion makes sense of the experimental results that, under quantum mechanics, are explained only by simultaneous occupation of separate locations, interaction from a distance and by no physical means, backward-in-time causation, and so forth.

The assumption that Dr. Little replaced – the assumption about wave direction – affects none of the math that quantum physicists have used to predict the results of experiments. The math works the same, regardless of wave direction. It supports Dr. Little's theory just as well as it supports quantum mechanics. But, unlike quantum mechanics, Dr. Little's theory obeys the rules of logic and hence may be right!

In fact, there is very good reason to believe that the Theory of Elementary Waves accurately describes reality. One test of a theory is whether it explains phenomena that it was not devised to explain. Little's theory does so in abundance, solving problems that involve Newton's physics, the Theory of Relativity, the inner workings of the atom, and even magnetism--another topic over which physicists have long parted ways with reason. The Theory of Elementary Waves in fact explains so much that it is far more than an alternative to quantum mechanics: it is a new explanation of physics, an explanation so elegant that Dr. Little shares it and its implications in a mere hundred and fifty pages.

Dr. Little has toppled a pyramid of mystical and impossible nonsense that less disciplined geniuses built over the course of a hundred years and which, for all that time, most physicists have admired. To replace that pyramid, Dr. Little has devised an ingenious theory that obeys the laws of reason and, so far, appears to be consistent with reality. His book offers fantastic encouragement for individuals who dare to think with discipline and persistence until they find answers that work."[95]

The bottom line here is that, as to risk analysis, our scientists cannot put the puzzle of risk analysis together, as to the LHC, or any other experiment involving risk in a manner deserving of trust, unless we can rely that said scientists are working with puzzle pieces of risk analysis which do not change shape. However, putting the puzzle pieces of quantum mechanics together, such as above illustrated in relation to the work of Dr. Little, is like trying to force the respective North poles of two magnets together; they are inherently opposed positions, which remain unreconciled, even if in contact. This fact of respectable, and wide, dispute as to baseline assumptions, very clearly undercuts the level of trust that the scientists *or we, who foot their bills and sometimes pay their dues*, can place in risk analysis based in quantum mechanics in complex systems. This difficulty in reliable predication is even more important when our world is at stake.

Chapter 21

With Many Worlds Physics, The LHC Risks The Multiverse

Overwhelming anecdotal evidence has shown the inability of humans, generally, to accurately predict outcomes in complex systems, and scientists, in particular, to accurately predict outcomes in complex systems. The various failures in prediction in the space program, such as the inch versus metric mistake made by scientists on a Mars mission, illustrate that Gaussian reasoning is not sufficient as a means of achieving an absolute rule out of risks. This is illustrated by the application of Chaos Theory to event prediction, and also by Taleb's work as to outliers.

The possible existence of other dimensions, from the unimaginably small (and unproven) Calabi-Yau manifolds[96], as brought to the popular reader by Dr. Randall, as necessary corollaries from String Theory, to the parallel dimensions postulated by Everett, Wolf, Kaku, and others in the Many Worlds hypothesis, have inherent within them many issues which affect proper risk analysis in the LHC context. Paramount amongst these, from my perspective, is that *we have no permission to harm other dimensions or the beings in them*.

An elegant example of Physicist Michio Kaku's eloquence is found at page 353 of his book; *Parallel Worlds*:

> "When we imagine the quantum multiverse, we are faced, as Trimble is in the story, with the possibility that, although our parallel selves living in different quantum universes may have precisely the same genetic code, at crucial junctures of life, our opportunities, our mentors, and our dreams may lead us down different paths, leading to different life histories and different destinies."[97]

From the Many Worlds approach, if Gaussian normal distribution based analysis has concluded that there were, for example, a one in a million chance of immolation of the Earth, then the LHC should be ethically impermissible on the basis that if that immolation occurred in even one of those million worlds; this would be genocide on a scale unprecedented in the history of man, the greatest crime ever to occur, since every living creature on those thereby affected parallel worlds, at least, would be killed. This reasoning applies to any Gaussian based risk analysis, as to the risks from the LHC, all before even getting to the possibility of outliers, as so well discussed by Taleb. That is to say, if there is any risk of planetary destruction, that destruction must occur in one or more of the parallel universes in the multiverse.

As discussed in the Prologue to this work, my early "hey buddy, wanna buy a watch?" level of contact with scientists, included a discussion of the trans-universal morality questions, in the light of possible trans-universal entanglement in the multiverse. I do not advocate the multiverse concept: The Quantum Physics paradigm is incomplete, and unconditional belief in it can sometimes become far more messianic than logical. However, for present purposes, since such belief is the dominant paradigm, we are justified in taking the Many Worlds hypothesis of Quantum Mechanics as an academic "given." The gravamen of this book is merely journalistic, the bringing together of well-established concepts which, in their comparison with one another show the value of caution. The single exception to a journalistic approach, the exception in which a personal evaluation of a physics issue is pitched, is in the trans-universal entanglement position stated in these essays, cumulatively published as *No Canary*: That if parallel worlds are accepted as an academically meritorious position, this in logic further requires a shared Big Bang, due to shared hyper compression of the precursors to matter, inevitably leads to the potential for trans-universal entanglement in the multiverse, once one accepts the multiverse in the first place. The rest of this is ordinary science-oriented lawyer work, only in this one instance is something potentially original mentioned, and, as to that, it stems from fair reliance upon what the involved scientists have said about their own works. As a starting place, as to the existence of parallel worlds, let us rely upon the explanation of multiple universes as explained by Dr. Ronald L. Mallett, in his poignant and academically popular book, Time Traveler:

"Quantum mechanics is the mechanics of sudden energy change. In quantum mechanics, energy cannot be gained or lost continuously, but only in fits and starts. In 1913, the Danish physicist Niels Bohr, often called "the father of quantum mechanics" showed that the electron orbiting the proton in the hydrogen atom can change its orbit only by either gaining or losing a certain definite amount of energy – no more and no less. These definite, or discrete, amounts of energy are called quanta of energy.

In 1957, physicist Hugh Everett III, then a recent graduate of Princeton, first applied quantum mechanics to the entire universe, which resulted in his many-worlds, or parallel-worlds, interpretation of quantum mechanics.

Quantum mechanics, in short, is a world of probability. In the ordinary, everyday world, when a pitcher throws a baseball, it is possible to describe exactly where the ball is and how it's moving. In the world of quantum mechanics, we can only say what will probably happen next, as we can't know exactly what an object is going to do.

In applying quantum mechanics to the whole universe, Everett found that whenever there is the possibility of more than one outcome for an event, there is a potential split in the universe. For example, suppose that at lunch you are trying to decide between a cheeseburger and a tuna sandwich. At the moment you make the decision, according to Everett, the universe splits into two parts. There is a universe in which you have chosen the cheeseburger, and there is also an equally real universe in which you have chosen the tuna sandwich. These new universes are parallel and separate. The you in the universe with the cheeseburger is not aware of the you in the separate universe with the tuna sandwich. Although this idea of a parallel universe seems incredible, it is completely consistent with the proven theory of quantum mechanics."[98]

At the beginning of *No Canary*, I mentioned that we sent the same letters to a bunch of physicists, and that the text of that July 11, 2008 letter would show up later. The letter was concerned with the moral implications of the possibility of the worst case event in the trans-universal entanglement matrix of the multiverse. The long letter to the scientists, summarizing positions on moral consequence, included that:

1.   First, a substantial body of mainstream scientific conclusion accepts the concept of parallel universes. In this regard, you are familiar with the work of Hugh Everett III and Michio Kaku and though from somewhat a different approach, the work of Dr. Lisa Randall and the greatly popular work of Brian Greene, the only one of these physicists I have heard in person. For instant purposes, I am reliant on the works of Kaku and Everett. The initial assumption to this quick thought experiment is that a substantial body of published physics work by serious physicists, both before and after the advent of String Theory in 1983, supports the existence of parallel universes; it is not a mere entertainment.

2.   Secondly, there exists, including the work for which the heroic Feynman achieved Nobel recognition, a substantial body of physics work by serious physicists, both prior to and after the advent of String Theory, which indicates that a near infinite, in any event some huge number, of rational possibilities will occur, as very difficult as this concept is both to imagine and to accept, in universes of differing degrees of adjacency from ours. By "differing degrees of adjacency" I imply that in a Gaussian sense, those close to mid-curve as we perceive it, will be more similar to one another, even from the standpoint of DNA identity, as posited by Kaku, and that those, to the extent that two dimensional illustration even makes sense here, farther away will have differing characteristics from the mid curve to a larger extent.

3. Third, we have the calculated odds by Sir Martin Rees, that the LHC has only a "one in fifty million" chance of causing the

destruction of the Earth. For purposes of this thought experiment, I posit and assume that this is a rational calculation of possibility.

4. Fourth, stemming from the above, I now assert that it necessarily follows from: 1) The existence of parallel words, and: 2) The accepted finding that all rational possibilities will occur in parallel universes, and: 3) That there is a one in fifty million chance of planetary destruction from the LHC, and: 4) That the number of parallel universes typically postulated as rational is far larger than fifty million, so that: The operation of the LHC must, therefore, must result in the destruction of one or more entire worlds (at least), filled with sentient and non-sentient life (regarding which there is an argument among scientists where lawyers fit in!).

As a second stage of this thought experiment is related to entanglement.

5. Entanglement, by which I refer to non-locality as first and best isolated by Bell's publication in 1963 is a real phenomena. As discussed by Nadeau and Kafatos; "After Bell published his theorem in 1964, a series of increasingly refined tests by many physicists of the predictions made in the theorem culminated in experiments by Alain Aspect and his team at the University of Paris - South. When the results of the Aspect experiments were published in 1982, the answers to Bell's questions were quite clear – quantum physics is a self-consistent theory and the character of physical reality as disclosed by quantum physics is non-local. In 1997, these same answers were provided by the results of twin-photon experiments carried out by Nicolus Gisin and his team at the University of Geneva."

6. The potential for entanglement between all matter exists because of the shared hyper proximity of the material immediately prior to the Big Bang. I have had the privilege of personally hearing Dr. Brian Greene lecture to this exact point, at Dominican University, in response to a question that I had asked about the Michelson Morely slit interferometer experiments, in the context of entanglement. I have composed an annotated 27 page essay on this (" No Canary in the Quanta", May 16, 2008), which I would be very honored if you would at some point read. It is focused on the LHC, and the outlier risks not comprehended by Gaussian risk analysis, in which case Taleb's work is central. I will send it along. That non-locality is "a property of the entire universe" and that "an undivided wholeness exists on the most primary and basic level in all aspects of physical reality" is discussed in the previously mentioned work of Nadeau and Kafatos (2001), with various supportive citations herein.

7. If parallel universes, vast in number, exist, which closely share traits of apparent reality, it must necessarily follow that these universes came from a simultaneous Big Bang event, essentially duplicate in timing and constituency. Regarding the proof of this

concept, I resort here Dr. Karl Popper's challenge, and invite falsification. It is beyond my ability to see a means of falsifying the contention just stated, that the Big Bang necessarily was the precursor to all parallel universes. I assume that others have seen this, but I'm a working lawyer, not a physicist, and have not yet been able to find a paper on it, though there must be dozens. The only mention I've so far found is in a New York Times article which is a discussion of the winning by Dr. Preskill and Dr. Thorne of a bet with Dr. Hawkins, the winning being based upon the supercomputer calculations by Dr. Matthew Chopuik (now in B.C., then in Austin). In that article, Browne writes, " But some weighty questions may be resolved by investigating naked singularities, including an explanation of the Big Bang, a naked singularity that is believed to have created our universe 15 billion years ago (and perhaps an infinity of other universes in other spaces and times." So, I stand with Popper, if you can falsify the position that the parallel universes started with a shared Big Bang, I would be respectfully interested in seeing your findings, to extent I could follow same, which might be limited.

8. It is respectfully asserted as the end product of the second stage of the thought experiment communicated with this letter that: If there was a shared Big Bang between parallel universes including our own, it must necessarily follow that there exists the potential for cross-universal entanglement between universes in the resulting multiverse. This is the best I can do. Hopefully somebody smarter can help with math. Here, again, I stand on Popper, falsify it if you are able.

9. Stemming from the preceding four statements concerning entanglement, and building upon the prior four integrated statements concerning the LHC, it necessarily follows that, via entanglement, destruction of a parallel world or, worse case (which seems unlikely due to gravitation issues) a parallel universe, must necessarily have some effect upon the universe which we populate.[99]

This problem, the ethical prohibition resulting from parallel universes, can be avoided by demonstrating that such parallel universes do not actually exist. Or, it could be empirically demonstrated that the CERN risk assessments were not probabilistic, as to strangelets, and mini-black holes, and if it could also be proven that there were no outliers, then the LHC could ethically proceed. Any doubt must be resolved in favor of life continuation in the multiverse. Disproof of parallel universes is not where modern String Theory currently takes us. Parallel universes cannot be dis-proven based on current academics, as Dr. Kaku observes in Parallel Worlds.

As stated by Kaku, "each distinct universe obeys Newtonian-like laws on the macroscopic scale."[100] Because the LHC, at the super-fine rat's tail of parallel possibilities   necessarily extinguish those "few" universes which represent the lowest ebb of probability, or, at least, a few Earths, within those rat tail universes. In substance, if there is any probabilistic risk at all, then at the very least an entire world of living sentient beings somewhat like our own, likely many of them, will be lost.

The LHC should be delayed while the issue of possible damage to alternative aspects of the multiverse is carefully evaluated. If the physics community allows the LHC to go forward without entirely and completely ruling out this issue, all the physicists in all of the surviving parallel universes, including this one, will be tainted with guilt beyond any guilt previously encountered in the human experience. The presentation of this issue to a particle physicist capable of voicing objection imposes upon that physicist to either: 1) disprove the many worlds hypothesis, or 2) empirically prove that there is absolutely no chance at all of damage to the Earth from the process, or 3) alert others so that the physics culture may make a decision as to its ethical obligations under this set of circumstances. Silence would be complicity in the results.

Chapter 22

A Slender Thread

So far as I can ascertain, prior to the summer of 2008 (when several tag along studies surfaced, during the late aspect of the Hawaii lawsuit controversy) there had been a total of three high-level published scientific studies within the prior decade which were concerned with the possible destruction of the Earth by the LHC or similar instrumentalities; and two of them are by CERN theorists, the second and latest was issued by CERN's LHC Safety Assessment Group and is called the "LSAG Report." These studies promising safety had been completed prior to the explosion that shut down the collider on September 10, 2008; one of the tag along studies, bearing an actual release date two weeks after that event, appears to specifically endorse the safety of the units which burst.

Let us, arguendo, concede these three studies as peer reviewed and assume their perfect academic regularity. What's wrong with this picture? Reliance upon academic papers issued by, or in direct conjunction with, the very entity whose dignity and survival is integrally related to the continuation of the LHC presents very obvious and substantial Conflict of Interest issues. For example, where UN teams are engaged in study of suspicious nuclear emplacements in countries of concern to that body, the teams don't rely upon the involved country's report, but insist upon independent verification. As Ronald W. Reagan said, "trust, but verify."

The small sampling of studies involved, are all from one community (yes, the physicists are more capable of the equations, though most are now computer dependent, but specialized familiarity does not impart wisdom). To consider the limited, non-experiment specific, non-independent studies involved as a sufficient safety analysis for a project which bears with it some possibility, however small, of the destruction of the Earth, is to hang our biosphere by an incredibly slender thread. It does not matter that the thread be silk, it is still too slender a thread upon which to hang the hopes and fate of not only mankind, but the other species on the planet, whose lives and DNA are being wagered by beings and processes which they cannot even vaguely understand. In this regard, I stand with the ants.

However, behind door two, is not the lady, but the tiger. Not only are two of the major studies paid for by CERN and written by CERN personnel, but, even more telling, these studies expressly contradict one another: The first being the source of the "1 in 50 million" number and, the second, despite the warnings of higher energy values at the LHC, issues a "no risk at all" conclusion.

Beyond those issues, conflict of interest, and the conflict between the studies, is there anything that we can point to in the studies themselves which gives pause to wonder as to the reliability of conclusion ? I am a mere lawyer, far from a physicist in specialty, but consider the following.

The most recent major study, which can be downloaded from the CERN website, is called the LSAG Report. It is an elegant, coherent, and rational analysis, and I view it, and its authors, with very great respect. The study should be read in its entirety, but some elements of it are clearly positions which stand upon their own merits. For example, concerning the black hole risk, the study concludes:

"One might nevertheless wonder what would happen if a stable microscopic black hole could be produced at the LHC [2]. However, we reiterate that this would require a violation of some of the basic principles of quantum mechanics – which is a cornerstone of the laws of Nature – in order for the black hole decay rate to be suppressed relative to its production rate, and/or of general Relativity – in order to suppress Hawking radiation."

With the deepest respect to all concerned, first, the above language seeks to sell the point that the "basic principles of quantum mechanics" are the "laws of Nature" and the word "Nature" is even capitalized to underlie the point, impliedly that "Nature" says that the LHC is okay, and "Nature" says that stable black holes will not occur, and that, of course, matter will not accrete to same. In fact, though, the "basic principles of quantum mechanics" are NOT (I can capitalize too) the "laws of Nature." First, if the "laws of Nature" were truly known, we wouldn't need the LHC to find them out. Second, "Nature" has been around way more than 13 billion years before Bohr, Heisenberg, and the Copenhagen Interpretation boys thought up Quantum Physics, only 80 or so years ago (and still controversial, like the song said, "after all these years"). Third, these supposed "laws of Nature," *quantum mechanics*, remember, have been described by Richard Feynman as "impossible" and "impossible to understand." These CERN physicists are not brighter than Richard Feynman, and with all due respect, and they do not understand Quantum Physics, because, as Feynman tells us, nobody does. It works, but how, like many medicines, the "why" remains unknown. Most importantly: There is now very severe, serious, logical, credible, and computationally unchallenged objection to the very underpinnings of Quantum Physics, referencing here the work of Lewis E. Little, as discussed in greater detail in the following chapter. Perhaps Dr. Little would cringe at my use of his work in this way, if so, I am sorry, I respect him vastly, and I haven't asked him about this. But his bottom line is clear, because his writing is clear and elegant, and the resulting application is therefore transparent: Not only is there no comprehensible proof that Quantum Physics are the "laws of Nature" even worse, far worse, as to the stability of black holes, there is deeply legitimate controversy, based, as one example out of many, from Dr. Little's work, as to whether the "basic principles of quantum mechanics" pitched now by CERN in this "laws of Nature" snow job, are even correct. It is not vital for us to know at this point, for LHC risk analysis, whether Dr. Little is right, or whether the LSAG Report is correct, on quantum mechanics being a "law of Nature." On this most crucial of all points, the stability of black holes, it is a sufficient showing of analytical instability. The credible underlying dispute exists, for this cuts to the unreliability of the LSAG Report conclusion, reliant, as it is, upon unconfirmed and controversial assertions.

Chapter 23

Thanks Anyway, But I'll Pass On The Magic

We have discussed the positions of various scientists, based upon the basic foundation that Quantum Physics is an empirically proven theory about the nature of existence. We have seen that, for just one example, the double-slit experiment defied a definition of matter, at least at the particle level, as either solid (as though there were such a thing) or wave.

The outcome of the double-slit experiment, and its kin, has led to many "spooky" conclusions. Quantum Enigma, by Rosenblum and Kuttner contains an excellent explanation of the often denied reality that, if one puts the Copenhagen Interpretation (sometimes thought of as FAPP - For All Practical Purposes) aside, FAPP being a mere "marriage of convenience" Quantum Physics leads us to some remarkably magical conclusions, "magical" conclusions which are accepted by the physics community as truth, even though Richard Feynman, supra, stated that "nobody can understand" quantum physics.

Call it full disclosure, I concede that there are persuasive factual data supporting that some phenomena which we currently define as paranormal, have in fact been shown to exist, including some relationship between consciousness and matter. For those interested in this aspect of physics, so long rejected by so many, the books by Lynn McTaggart, The Field and The Intention Experiment and the excellent work of Dr. Dean Radin, in Entangled Minds and The Conscious Universe, are great destinations, which survey the actual results of experiments and meta-surveys, persuasively showing something that challenges our basic "Western" viewpoint of utter separateness between the self and its surrounds. There are, no doubt about it, also many junk science books cluttering the shelves in this field. But there are also works of solid science, and also journalism, such as The Holographic Universe, by Michael Talbot, surveying, in part, the "hidden variable" work of physicist David Boehm, and there is Rupert Sheldrake's work regarding the apparent presence of "morphic fields" (discussed, infra) in *The Presence Of The Past*. These include books which employ solid logic and which report on very strong experimental data showing that there is more to the world that we sometimes call "psychic" than meets the eye blinded by prejudice from seeing that which the experimental data actually show. Currently, the fashion is to lay these anomalies at the altar of Quantum Physics. Yet, it is also possible that there is some other basis for the findings, for one example, of the Princeton PEAR project, so recently closed after almost three decades of solid science, sneered at only by those who have not read the studies and done the math.

At least rational analysis must concede that the reliability of Quantum Mechanics for predictive purposes in complex systems is questionable. One reason for this is that the groundbreaking work of Dr. Lewis E. Little cuts strongly against the most basic foundations of Quantum Mechanics and, in doing so, presents a cogent alternative to many "Quantum Weirdness" scenarios which have been viewed by some authors, both credible and incredible, as an explanation for at least some of these phenomena. To say that Dr. Little's word may undercut Quantum Physics as an explanation for these phenomena is entirely different from challenging the experimentally proven fact that such phenomena, that some such phenomena are experimentally proven, such as proven at the Princeton PEAR Project, and as discussed, for example, in the works of Radin and McTaggart.

Yet, though they be real, the "why" of these phenomena remains largely unknown. The case for Quantum Physics in this, and in so many practical regards, is very strong, particularly as to Entanglement, which we must, I feel, regard as now proven with certitude. It is not necessary to *No Canary* to disprove the basics foundations of Quantum Mechanics, rather, the critical point is that the serious questions about the reliability of predictions based on Quantum Physics are directly relevant to the thesis of this book, namely that we lack sufficient certitude to justify turning on the LCH.

The work of Dr. Lewis E. Little provides strong theoretical challenges to the foundations upon which the scientists who thought up the LHC subscribe.

First, ardent defense of String Theory necessarily involves zealous belief in something which lacks empirical proof, and is therefore faith-based reasoning: We should not be risking the Earth on the basis of religious conviction, including the religion of science. As Kuttner and Rosenblum point out, quantum physics has always, despite the formidable mathematical proofs and practical applications, nonetheless been a source of unresolved enigma. It has never been clear to any published scientist why, fully, quantum physics works, though it is very clear that the equations upon which it is founded work very well indeed, and that the resulting practical applications have worked well too. On this proof, based upon reasoning back from results (as opposed to forward from empirical data), quantum physics itself has become an untouchable grail. And yet, very respectable credentials now argue that, as to quantum physics as a whole, and in particular including the dual slit experiment, related experiments on particle migration, that, like Gershwin said, "it ain't necessarily so."

Dr. Lewis E. Little received his BA in physics from Brown University in 1962, graduating with highest honors. He received his MA in physics from Princeton University in 1965 and his Ph.D. in physics from New York University in 1974. I am a great fan of Amazon, Kindle and all. Wherever else these positive reviews may have been published, the voices respectfully endorsing Dr. Little's at Amazon, which you can see this minute, show that Dr. Little is widely respected.

For example, Frank Schneider, Ph.D., of Jet Propulsion Labs endorsed that: "Physicists will marvel at the way Dr. Little is able to penetrate to the crux of the matter with crystal clear explanations on so many related subjects in so short a space....A good read even for experienced professionals."[101]

Archie McKerrell, Ph.D., a Theoretical Physicist from the University of Liverpool said: "I recommend this important work. Little's theory makes quantum physics understandable."[102]

Most germane to the matters treated here, Michael Flagg, a Nuclear Engineer at the University of Missouri Research Reactor Center has added his review that: "Like Copernicus demolishing the cycles and epicycles of Ptolemaic astronomy, the Theory of Elementary Waves has the potential to sweep away decades of absurdity which have grown up on the basic observations of quantum mechanics."[103]

In the foreword to The Theory of Elementary Waves, Robert R. Prechter, Jr. says that Dr. Little "had a vision as revolutionary as that of Copernicus 350 years earlier"[104] and that "he not only revolutionizes the fundamentals of sub-atomic physics but also reclaims the fundamentals of scientific philosophy."[105]

Relevant to the intellectual tribalism earlier here described, at the commencement of his Forward to Dr. Little's *The Theory of Elementary Waves*, Robert Prechter Jr. discusses "error pyramids" and cites the epoch in science before Copernicus in example:

"My favorite books are ones that dispel decades, centuries, or millennia of misconception in one brilliant stroke. Intricate, long-standing theories built upon such misconceptions are called "error pyramids." One was built around the false premise of a geocentric universe. Some planets appeared to act weirdly as a result of this error, such as Mercury moving "retrograde" with respect to the Earth. Astronomers produced complicated calculations and theories to account for the motion of the planets in a perceived geocentric solar system, even to the point of perfect prediction. Yet, Copernicus' 1543 book on celestial mechanics, On the Revolution of the Heavenly Spheres wiped away the entire vision and replaced it with one that made sense not only in the calculations, but also, for the first time, in the physics.

This is precisely Lewis Little's achievement with respect to the behavior of sub-atomic particles: making the physics fit the calculations."[106]

The Theory of Elementary Waves: A New Explanation of Fundamental Physics, by Dr. Lewis E. Little[107] challenges the very roots of quantum physics. Not only do we have a great hero of science, Dr. Feynman, stating that "nobody understands" quantum physics, we have generations of physics students being asked to commit to their world view phenomena that their teachers, if honest, will acknowledge, as Dr. Feynman also said "are impossible, absolutely impossible" to explain based on human experience.

Examples of this "impossible" behavior include many contradictions, including that elementary particles simultaneously exhibit the properties and behavior of both particles and waves, a notion which led to the claim that one particle can be in two places at once. As Dr. Little points out, the links in this chain of absurdity have led to bizarre extremes, such as the well-known physicist who wrote in utter seriousness that "the moon is demonstrably not there when nobody looks."[108]

Dr. Little's book contains a comprehensive listing of the "mysteries" which currently comprise the "mystery school" of quantum physics. Since the double slit experiment is at the core, or tap root, of quantum physics, it is appropriate that Dr. Little has faced up to that enigma first.

Dr. Little, again, a Ph.D. physicist of high background and good standing, proposes that:

"Elementary waves are present at all times throughout all space in the universe. A detector or other object from which elementary waves emanate does not actually create or emit those waves. Instead, the sub-atomic particles that make up a detector act to rearrange, or 'organize,' the waves impinging upon them. No net 'quantity of wave,' is either created or destroyed in the process.

Waves interact with subatomic particles as they pass by in the immediate neighborhood of the particles. Some waves interact directly with a particle. Other waves interact with waves that have already interacted with the particle.

Waves' interaction with a particle serve to 'scatter,' the waves in all directions come out from the particle. Many of the scattered waves will carry a common phase and a common 'marker,' the common phase and marker constitute the 'organization,' of these waves. We will speak of waves originating at a detector, but it is

98

only the organization, not the waves themselves, that a detector originates."[109]

Perhaps a clue to the underlying consistency which Dr. Little's theory seeks to show us can be found from a heretofore unexpected source. Perhaps some readers will agree the possible intellectual overlap which I see between Dr. Little's work and the writings of Dr. Rupert Sheldrake, in *The Presence of the Past*, where Dr. Sheldrake discusses the symmetry in snow flake crystals as shaped as a result of a lattice morphic field, and, in my view at least, a similarity with Dr. Little's Elementary Wave conclusion, in that theorganization of the material object is influenced by an underlying "field," which necessarily, I believe and here assert, implies a wave aspect. Just as, with humility, Dr. Little's work is endorsed, so I also recommend Dr. Sheldrake's *The Presence Of The Past* and further that the discussion of crystal structure here mentioned is one of the most interesting discriptive aspects of that work.[110]

Later in, Sheldrake discusses such fields in the context of formative causation[111], and, at least in my view of it, takes the position that in more complex systems, the higher levels of complexity are derivative from underlying field structure.

Sheldrake's morphic field hypothesis is of interest here for multiple reasons. First, it is flat-out interesting to note that Sheldrake has proposed a system of fields which may relate to the Elementary Wave fields that Dr. Little proposes as a less romantic, yet also far less weird, explanation for phenomena which are otherwise now pigeon holed as being quantum artifacts. If this overlap is present (I have not found it elsewhere discussed) then these two views may have a mutually reinforcing intellectual framework, whether the respective authors thereof like it or not, an unknown. In turn, this is of interest to our LHC discussion, because Sheldrake's work is one further indicator of the presence of hidden variables (here intentionally using the term as employed by David Boehm) which are not, and as yet cannot be, part of the LHC risk calculus, a further increment of indication that there is insufficient consensus in the physics community to allow such a potentially draconian experiment to go forward.

Readers will recall that in our earlier discussion of the double slit experiment, it was the "wave-like" signature at the target plate, emblematic of an interference pattern, which led to the conclusion that the entering particle acted both as a waves, and as a particles. My own acceptance of this very attractive outlook, based upon the experimental data, an acceptance now challenged, led me to personally prefer the term "wavicle."

The outcome of the double slit experiment, and many related experiments such as the slit interferometer experiments of H. Kaiser, et al., as discussed in Dr. Little's book[112] have also led to such strongly held beliefs as that the particle may be in two places at once, until it becomes corporeal at one spot or another, and quantum physics experiments which have been viewed as supporting that a given "effect" may occasion its "cause" earlier in time. In fact there are many mysterious aspects of quantum physics, so odd in their nature that they were items of faith for the faithful, at least until Dr. Little's recent publications showing at least the possibility of legitimate alternative explanation.

Dr. Little's book includes a full critique of quantum theory, including Heisenberg's Uncertainty Principle, Bell's Theorem, and the "double-slit" experiment. He provides consistency of theory with Einsteinian Relativity. It is, in essence a tour de force indictment of the widely held belief in the "magical" nature of quantum physics, includes very detailed discussions of atomic decay, and offers potential new insights in a wide variety of fields, including biology. Sheldrake's work, on the other hand, while supportive as to the existence of, his term, The Primal Field of Nature, may both re-enforce the Elementary Wave concept, and, along with works such as those of McTaggart, point to an underlying "zero point field" not as yet well comprehended within generally accepted public consumption physics paradigms, the modulation of which (in this sense reminiscent of Koestler's speculation as to meson modulation) may be a mechanism of experimentally verified phenomena, such as the pre-attack Random Number Generator anomalies which occurred on September 11, 2001, as described by, for one, Radin in *Entangled Minds*[113] Thus, the abandonment of all or part of the quantum paradigm set does not require the abandonment of, for example, experimental data showing entanglement, rather, another means are suggested.

It is not the outcome of these dialogues, however, which is crucial to our discussion of the LHC. The crucial part is that these vast chasms in viewpoint exist within the physics community at all, since their existence cut to the very most rock bottom foundations of all physics reasoning. The field is in great disarray. Persons or teams lacking cohesiveness of baseline principle should not be trusted with the operation of dangerous instrumentalities, let alone those capable of destroying our biosphere.

Seldom, perhaps never since Darwin, has there existed a starker contrast in world views. Dr. Little's provision of a comprehensive response to Quantum Physics stands in poised opposition to such excellent publications on the subject such as that discussed by Kuttner and Rosenblum in Quantum Enigma. Currently, this is the most vast of the chasms separating one physicist from another, as different members of the physics community adhere, in their intellectual tribalism, to points of view which are not only different, but also mutually exclusive of one another.

Given this disarray, what are the factoids, and what are the facts? In fact, we each and all, tend to see reality via the tint of the lens of view. A close and trusted friend of mine, Patrick R., who is senior litigation partner for one of the most prestigious law firms in the United States, speaking of the testimony of experts, has said to me: "impartiality is a myth." If one is a String Theorist, then it all makes sense in the universe of super-strings, and the many dimensions are then contrived by mathematical physicists to buttress that position.

There are activities in human affairs which require extremely high certainty. Thus, for example, Army parachutists are skilled in the packing of their own chutes, and for obvious reasons, only those who have that skill are allowed to engage that process. Only carefully trained pilots are allowed to operate airliners. A young person being taught air rifle or bow and arrow safety is told never to shoot without a backstop. General Aviation pilots are taught to visually check the amount of fuel in the tanks, and not to rely totally upon a gauge. We don't have to be absolutely sure of the gram weight of a tennis ball, but a competent pharmacist must be sure, down to the milligram, when dispensing most medicines, since they all have the potential for unanticipated side effects.

In international relations, including as to war, we have learned the very sterile phrase "collateral damage" to sugar coat the terrible tragedies that come with the imprecision of combat, but still, in some circumstances (such as stopping Hitler), we persist, because though we know that there will be unanticipated consequences, the stakes are perceived as so high as to justify those risks, and those outcomes. That never diminishes the tragedy for the persons wrongly affected, but at least there are, modernly, attempts to abate the consequences of mistake, such as the provision of medical and other aid to families harmed as "collateral damage" by air attacks which did not go precisely as planned. Or worse, when they do, in which case Robert MacNamara's comments on the fire bombing of Tokyo, as statistician for General Curtis LeMay, show that just as sorrow can be a permanent stain, so can regret. However, while solace or rebuilding can be provided to a nation or a people, there is no way to provide assistance to the gravitationally compacted ashes of a destroyed Earth.

After reading many dozens of books relating to quantum physics, I feel that Quantum Enigma by University of California (Santa Cruz) physics professors Bruce Rosenblum and Fred Kuttner is the most important book currently available about the limitations which have always been inherent in The Copenhagen Interpretation. As with the works of Dr. Dean Radin, Rosenblum and Kuttner empirically show that there remains an unresolved enigma about quantum physics, including the apparent role of consciousness as a factor influencing the nature of physical reality. Nowhere is there a proven "theory of everything" despite many attempts: My most recent personal reading in this regard is *The God Theory* by astrophysicist Bernard Haisch. Now, with Dr. Little, we have a further, highly qualified, non-faith-based, logically consistent, and mathematically consistent viewpoint which challenges the very cornerstone of physics dogma. There is no consensus in the house of physics as to what composes its foundation, it is a house composed of the competing business cards of its creators.

Empirically, there is a great deal that we don't know. In particular, we don't have a back drop to catch the bullet aimed at life on Earth if these physicists miss their mark. To go forward with this LHC, with so many unknowns, is a reckless act, imbued with the potential for a catastrophe beyond any in prior history.

Finally we cannot know what consequences we are unable to see, but which may be visible to small elements within the scientific community who are so blinded by obsession that they are willing to proceed despite knowledge of a known risk to many. Until there is democratic methodological approval of the taking of such a risk to the planet, the experiment should not be allowed.

Chapter 24

A Circumstantial Case

While I had a strong view of the LHC subject area, as shown by the stolen hours of work in 2008, it was not until the evidence piled up in late Summer of 2009 that I truly understood how limited the LHC safety analysis had been. The 27 page paper circulated in May of 2008, and the letters to scientists sent at the same time, did not result from emotion, but rather were work undertaken, as I would undertake a case, with a detached and perhaps mechanical approach, an analysis undertaken out of a sense of duty.

It wasn't until deep reading in the Summer and Fall of 2009 that I realized that, to use an appropriate legal term, the defense of the LHC was entirely a matter of circumstantial evidence. Thus it was only in that mid-September, that a sense of personalized urgency came into this work for me, along with the epiphany that I had to start approaching this as the most important Trial Brief of my long career. So, though I attended to my practice obligations, and worked at being a good dad, and tried to be a good husband in the midst of it, it was in September that I started to crack the whip on myself, and on our team approach to this, project.

None of this is important, of course, it is just my personal story, and I am not a VIP and never will be. But I mention the calendar of my consciousness, because a side effect of this timing was that, by luck or synchronism, it was at this time of intensity that a few articles of great pertinence also happened to be published, including the story of Cameron Todd Willingham, which was extensively covered in the second September 2009 edition of *New Yorker* magazine, in an excellent article by David Grann titled "Trial By Fire"[114].

The Willingham case provides a fair and useful example of the importance of circumstantial evidence to the operation of our system of justice, the crucible in which we figure out our most difficult and tragic problems, and for reasons which shall be seen, Cameron's story also illustrates problems in the standards of proof relied upon in defense of the supposed safety of the LHC.

The prosecution of Cameron Willingham would have been impossible had it not been for the use of expert testimony based upon circumstantial evidence, as presented by experts in forensics, to prove at trial that Willingham had in fact killed his own three daughters, by setting fire to the home in which they were sleeping, and in which, so the jury was led to finally figure out, Willingham had been pretending to sleep.

The tragedy took place in December of 1991, though one could argue that the full extent of the evidence, including a stove as the more likely cause of the fire, did not reach broad and public light until the Fall of 2009, with the publication of the New Yorker article.

On behalf of the prosecution, expert testimony was able to establish a solid case, including the testimony from deeply experienced arson investigators that: 1) Indicators of "puddling" proved that an accelerant was used: 2) Lab science established that highly flammable "mineral spirits" were without doubt in stains at the front doorway: 3) A "craze" pattern in the windows showed that an accelerant had been used, since the crazing is caused by the higher temperatures caused by accelerants, and: 4) Expert from a psychologist focused on a detailed image on the wall demonstrated that the defendant was obsessed the combination of fire and death, and had finally acted out on this obsession, and: 5) The guilt of the defendant was further demonstrated by testimony from a cellmate, directly establishing from his own words that Willingham had killed his own kids (because, as the local District Attorney put it, "The children were interfering with his beer drinking and dart throwing"), and: 6) Char and smoke patterns on the walls, ceiling, and one of the beds proved that the fire was hotter below than above, which could only have happened if a flammable liquid was used. Many other strong evidentiary points were made by the prosecution experts, and readers of No Canary are strongly urged to read the full New Yorker article, using the link at the end of this chapter, I cannot do it justice here, and it is an important article.

Our concern is more limited in scope, as a window to the values and problems which go along with our system's use, often essential, of Circumstantial Evidence. So far as I know, Circumstantial Evidence is allowed in all American jurisdictions; I believe that a fair and useful portrait of it can be seen by simply using our California Jury Instruction on the subject, similar in function to that found for criminal trials in all States:

"223. Direct and Circumstantial Evidence : Defined

Facts may be proved by direct or circumstantial evidence or by a combination of both. Direct Evidence can prove a fact by itself. For example, if a witness testifies that he saw it raining outside before he came into the courthouse, that testimony is direct evidence that it was raining. Circumstantial Evidence may also be called indirect evidence. Circumstantial evidence does not directly prove the fact to be decided, but is evidence of another fact or group of facts from which you may logically and reasonably conclude the truth of the fact in question. For example, if a witness testifies that he saw someone coming inside wearing a raincoat covered with drops of water, that testimony is circumstantial evidence because it may support a conclusion that it was raining outside.

Both direct and circumstantial evidence are acceptable types of evidence to prove or disprove the elements of a charge, including intent and mental state and acts necessary to a conviction, and neither is necessarily more reliable than the other. You must decide whether a fact in issue has been proved based on all the evidence."[115]

After providing the jury with the definitions of direct and circumstantial evidence, the law in California also provides guidelines, in another Jury Instruction, about, as to direct and circumstantial evidence, what is sufficient, and what is not:

"224. Circumstantial Evidence: Sufficiency of Evidence

Before you may rely on circumstantial evidence to conclude that a fact necessary to find the defendant guilty has been proved, you must be convinced that the People have proved each fact essential to that conclusion beyond a reasonable doubt.

Also, before you may rely on circumstantial evidence to find the defendant guilty, you must be convinced that the only reasonable conclusion supported by the circumstantial evidence is that the defendant is guilty. If you can draw two or more reasonable conclusions from the circumstantial evidence, and one of those points to innocence and another to guilt, you must accept the one that points to innocence. However, when you are considering circumstantial evidence, you must accept only reasonable conclusions and reject any that are unreasonable."[116]

There was no direct testimony that Cameron Todd Willingham had murdered his children, as nobody had seen him do it. But there was circumstantial evidence based in physical findings at the scene that were convincing as to his guilt. The jury convicted after a two day trial, in which his appointed general practice lawyers told him not to testify on his own behalf, despite the damning circumstantial and opinion case that the prosecution established. In fairness, criminal defendants seldom take the stand in major cases, though in this instance, rebuttal based upon apparent character may have been the only card I defense counsels' deck: I don't know, not having been there, I do not criticize the actions of counsel, but it is fair to say that they were not forensic specialists.

So, as *The New Yorker* article reports, Cameron Todd Willingham was sent off to state prison for the murder of his children, "my babies" as he called them, and was branded a sociopath, the jury haven taken his skull tattoos and death by fire posters into account, after an psychologist expert said that they were relevant to his motivation. Yet, despite Cameron's conviction, based on circumstantial evidence, the good news is that his reputation was saved when his innocence was eventually compellingly established, long after the trial and all the appeals, truly dedicated scientific analysis, provided by inventor and doctoral chemist Dr. Gerald Hurst, who, having invented the Mylar balloon and other useful products, had the time to devote to determining the truth in arson cases, where his work has had a major impact on the field of arson investigation. Briefly (again, please see the excellent article), Dr. Hurst compellingly established, after months of work, in a report submitted to the proper authorities, that: 1) The prosecution testimony that the fire had burned "fast and hot" due to the hotter burning of accelerants was flatly wrong, since experimental evidence has clearly established that accelerants do not burn hotter than wood, and: 2) The crazing of the glass was more likely by rapid cool down from sprayed firefighter water, and: 3) As to the speed of the fire, the "puddle effects" as well as the high speed, were clearly tied to a stage in fire progression called "post-flashover" and those elements of physical evidence, used as circumstantial evidence of guilt, were not in reality indicators of accelerants, and: 4) The "mineral spirits" were which had been identified were innocent, because the family BBQ was on the front porch, where the samples had been taken. As the article described Dr. Hurst's findings regarding prosecution "expert" Vasquez:

> "Hurst was also struck by Vasquez claim that the Willingham blaze had 'burned fast and hot' because of a liquid accelerant. The notion that a flammable or combustable liquid causes flames to reach higher temperatures has been repeated in court by arson sleuths for decades. Yet, the theory was nonsense: experiments have proved that wood and gasoline-fueled fires burn at essentially the same temperature."[117]

A link to this twenty-page article may be found on the Innocence Project website, which will direct you to the article download link on the New Yorker website. What about the "skulls and death" images on the defendant's walls? The object in question was simply an Iron Maiden poster. The testifying cell mate ? A heavily medicated robbery convict, he later recanted his testimony, and then still later recanted his recantation. A toppled heater is far more likely to have caused this fire than any action of this apparently innocent father.

Finally, scientific evidence proved this defendant's innocence and was, just four days prior to his scheduled execution, sent to each of the fifteen members of the Texas Board of Pardons and Paroles, which issued their vote prior to the scheduled date with death. They denied the petition. The State of Texas then killed Cameron Willingham by lethal injection.

The New Yorker article references a Texas appellate judge who called the Texas clemency system "a legal fiction." As the New Yorker article then continues:

"The Innocence Project obtained, through the Freedom of Information Act, all the records from the governor's office and the board pertaining to Hurst's report. "The documents show that they received the report, but neither office has any record of anyone acknowledging it, taking note of its significance, responding to it, or calling any attention to it within the government, Barry Shenck said."[118]

Of course, the story above was about just one innocent man killed by the state government in Texas, but a comment obtained during his investigation by the article's author, David Grann, illustrates the tendency of participants, including witnesses, to report that which they believe will please the listener:

"Dozens of studies have shown that witnesses' memories of events often change when they are supplied with new contextual information. Itiel Dror, a cognitive psychologist who has done extensive research on eyewitness and expert testimony in criminal investigations told me, "The mind is not a passive machine. Once you believe in something – once you expect something – it changes the way you perceive information and the way your memory recalls it."[119]

What does this sad story have to do with the LHC ? The CERN study which led to the "1 in 50 million" statistic, relied upon by Martin Rees, and then repeated as gospel all over the Internet, was based on circumstantial, or "indirect" evidence. In the law, in working before both Judge and Jury, we seek through forensic means an explanation for the past, and we sometimes have the benefit of both Direct and Circumstantial Evidence. While respecting its value in the law, history, such as through the work of the Innocence Project, has shown that even carefully studied circumstantial evidence can lead the finder of fact into tragic error, as Cameron Willingham's life and death has shown us. It is respectfully suggested that Circumstantial Evidence alone is not a sufficient basis upon which to risk our entire planet.

Chapter 25

A Modern Defense of Democracy

Will the physics community take any of the points we've explored to heart ?

There are a few points on which we should all be able to agree, the most important of these being a point upon which both critics of the LHC and many of the experiment's staunchest supporters can agree: A possibility exists that this experiment may destroy the Earth. The position of many in the physics community, as we have seen, is not that destruction of the Earth by the LHC is strictly impossible, but rather, so frequently encountered reasoning goes, the chances of a worst case outcome are so tiny that we, taking a page from that sage of sages, Alfred E. Newman, have nothing to worry about.

So very frequently, the assurances from the physics community come couched in condemnation. Sometimes it is silly: "There's a possibility, after all, that you might vaporize while shaving today, but you don't actually worry about that, do you?" Where is, I wonder, this phantom shaver ?

I've run my criticisms of LHC safety past a few friends with varying degrees of professional background related to physics. One encounters the "don't worry about vaporizing while shaving" level of dialogue very frequently, in person, and as seen on the Internet. What I have not ever encountered are "drill-downs" to the very most basic and serious questions treated in this book, including a small number of studies and the frailty of their application.

The fact that such a possibility exists at all is what brings this whole problem set into undiscovered country for mankind. The territory ahead is uncharted; no worlds have gone there before, so far as we know. This is a level of possible bad outcome which is unprecedented in the whole history of humankind. That is why the issues deserve to be examined very carefully. Even were we to decide, after study, that this new knowledge, about which the physicists obsess, were worth risking all life, at least there should be a few independent studies, specific to the LHC, before pushing any buttons which might give new meaning to the phrase, "lights out."

This book has never been disrespectful of science; to the exact contrary, the insistence here has been to suggest that we, the people of the world, each and all affected by this, should follow the path of studied logic, with emphasis on the word "studied" wherever the resulting data may take us. Our path here, as to our own perceived universe, has not been to reject the LHC, but to insist that it be empirically evaluated on the basis of direct evidence, and not switched on for the planned extended periods, on the basis of circumstantial evidence. This need not be a position without compromise; though clearly there should be truly independent evaluation of risk, prior to ignition, a compromise approach may be to allow the machine on, but for a far lesser, certain, and measured period of time. The physicists involved with the project already point to the overwhelming flood of data which it may provide; if this be the case, then study the trickle first, for within even a narrow band of experimentation, much, perhaps alas, may be learned. Yet, again, proponents and objectors alike, cannot we not at least concur that these are very big questions.

The biggest questions of all, as we've seen have centered around whether people, experts in particular, can in fact be objective. A dear friend of mine, the senior litigation partner for one of the nations most prestigious law firms, has said to me, as to expert witnesses, "objectivity is impossible." We put on a show about it, but those of us "in the know" from years of trial practice, often (such as in complex medical cases) involving tremendously complex data, realize that the bias of the observer is more typically than not impossible to entirely filter out. For one example, those experts making their living in the provision of even handed scholarly evidence for medical malpractice defendants would not be repeatedly hired by defense counsel, if they'd developed a habit of always finding fault with the doctor's side of the case.

So, we've seen here that persons of bias, more often than not cloaking themselves in the boxer's robe with "objectivity" stenciled on the back, clump together in their views, not dissuaded by even the most scientific commentary to the contrary of their "approved" positions. This is the nature of humanity, that we are from the womb to dogma borne. But, this never mattered before, as much as it does now.

We have seen the rages of obsession that go with enthusiastic "bubbles" like the Tulipmania bubble, the dot-com bubble, and most recently, the credit-default-swap bubble. We have seen, as to String Theory, for one example, that there are serious concerns, voiced by reputable scientists, as to whether or not a new religion is being formed, since String Theory, empirically unproven, is defended, and its detractors put down, with a zeal that would do a cheerleader proud. We have also seen that those with a history of cheerleading may not be entirely ready for larger and tougher problems, however fundamentally decent they may be inside.

Should ordinary people have a role in deciding our own fate, where the LHC is concerned? Those who have brought the LHC experiment this far have done so in an anti-democratic manner. Dr. Lisa Randall suggests that science works best using a "top down" approach, where those who "know best" can impose their reasoned will upon we, the huddled masses. And where are the voices of the scientists and physicists who believe this experiment is too risky, or the voices of the many educated and intelligent non-physicists - do we have a say in this?

One question that arises from this controversial and, dare I say, scary, experiment is this: Should Democracy have a voice in science? I understand the idea that some common folks "can't do the math" and that some complex decisions must be made by those best educated in certain fields. Yet, it is our birthright to protect ourselves and our family and fellow citizens. We have earned the right to read the scientific studies explaining the risks of this potentially deadly experiment. Most governments require extensive Environmental Impact studies to build a single building. Shouldn't we give as much attention to studying the effects of an experiment that threatens our entire planet Earth? Democracy seeks a working and fair balance between the individual and the State. I believe all people are entitled to be informed and reassured of the safety of the actions of those who claim to lead.

So, personal liberty is an important reason for democracy. Historically, we would look, amongst many others (the Anti-Federalist Papers, for example) to the work of John Locke in this regard. We need only think of the worst dictators of the last century to see the value of democracy in terms of values for the people.

Democracy is always "in trouble." In America, our democracy arose out of The Enlightenment, which, while held back by legacy forces in Europe, blossomed in the New World, and was codified in The Constitution of the United States.

Yet, some at the highest realms, sometimes visible, sometimes marching to drummers unseen in celebration of soldiers long past, clearly long for a return to more Feudal days, so that they might be once again treated as hereditary nobility, seeking a return to a top down approach, not only as to science, but as to the overall management of the affairs of our global humankind. This tension of viewpoints will never stop, and yet it is at the fulcrum of balanced liberty that the greatest leverage of humanity is most efficiently and creatively applied.

It is an oddity of our system, as it has most recently operated, that those who have been least challenged by the difficulties of livelihood, and thus have the least experience in facing the tough decisions of life, death, risk, and making a payroll, have been placed in positions of the greatest possible power. There is no need for me to get personal here, those who read the papers with eyes to see have witnessed a hurricane of excess coming from entrustment of great responsibility to the hands of persons not honed by the experience of struggle. Malcolm Gladwell, in *Outliers*[120] has pointed out that an incredible constellation of privileged synchronism was has laid the bedrock for many great successes which resulted more from the intertwine of privilege than the education provided by struggle. For pertinent example, in the Hurricane Katrina situation, the failures of FEMA's top leadership remain legendary and remind us of the risks inherent in putting people in authority who have not been equipped for its use. To belong to a large scientific organization is to recognize that specialization is so intense that the big perspective of the forest is entirely lost to those preoccupied with a single tree. We have all seen many examples in recent decades of a decline in the value assigned to the rule of the people, and a commensurate increase in the respect shown for "top down" rule.

B.F. Skinner, a behaviorist who influenced American thought in the 70's, and whose influence has affected government and academia ever since, argued that we should not be concerned about the "voice of the people" and should rather rely upon the skills of "Controllers" to look after civilization, as Skinner phrased it his book, *Beyond Freedom and Dignity*.

Fair enough, within the great mass of people, many believe in silly things. Yet, in sharp counter-point, both history and current events are studded with examples where narcissistic rulers have ruined many lives. Democracy is the answer to that, and should be part of the balanced process through which the words "science" and "ethics" are brought into intertwine. It is the best answer we have, and this is not just a matter of philosophy, it is a matter of empirical proof.

In political philosophy, the answer to those who favored the feudal approach (inevitably, of course, those who were comforted by it) was Rousseau, who posited the idea of The General Will, which said that the informed decision of the largest number of people, constituted The General Will, which would produce the best decisions for civilization. We have all seen the damage that can occur when too much power goes to too few hands. But is this just a romantic ideal, or is there some basis for it in reasoned thought, and maybe even science?

Michael J. Mauboussian is a respected Wall Street scholar and investor and an adjunct professor at Columbia Business School. As discussed in a March 11, 2006 *New York Times* article called "The Future Divined By the Crowd"[121], every year he has a group of students predict who will win in major categories of the Oscar nominations. In 2005, their pick was right three fourths of the time. That by itself is amazing, since there are so many contenders for each category. But there is a more important insight buried in that data, which is that the group always did better, in predicting outcomes, than any individual in it. As the *New York Times* article states, out of the 47 students in the group, only one matched the accuracy of the consensus. And as the article further explains, referencing the Iowa Electronic Market, run by the University of Iowa: "The consensus almost always beats the polling data." Thus, Rousseau, with General Will, "got it right" and that majority determined predictive decisions will strongly tend to be better than the decisions of any one individual.

Thus, when we defend democracy, we are not "just" defending the rights of individual citizens, we are defending the quality of decisions made by the civilization itself. Democracy is not just good for the individual, it is crucial to the society.

In this case, the need for democratic is justified not only in the interests of the many, but, since the lives of all are at stake, also in the interest of the few. We have seen from clear data that the allegations made by CERN et al about the safety of the LHC have been conjectural opinions, based on disputed modalities applied to circumstantial evidence, and influenced by faith-based reasoning. It is time, it is the eleventh hour, to say to these scientists that they may not use our entire world for the laboratory pursuit of their obsessions.

Harry V. Lehmann, November 5, 2009

About the Author

Harry Vere Lehmann attended public schools in his hometown of Novato, California, and then attended the College of Marin, in Kentfield, California, where he served as president of the student body and president of the graduating class of 1968. Next attending San Francisco State, now CSU San Francisco, arriving in the aftermath of the worst student riots there, Mr. Lehmann was elected as president of the student body on an anti-violence platform called Satyagraha, after M. Gandhi, and served in that office in the 1969-1970 years, seeking avoidance of violence by all sides. Indelibly angered at the Gulf of Tonkin Resolution, as pushed through the U. S. Senate by President Johnson, in 1970 Mr. Lehmann accepted the invitation of then California Governor Ronald W. Reagan, to run the aspect of his re-election campaign concerned with college and university voters. In 1971, while attending law school at the University of San Francisco, Mr. Lehmann next served as a Sierra Club National Law Intern, with specialty focus on the sought formation of the Golden Gate National Recreation Area, which was soon thereafter funded as a result of a large team effort. As a "Liberal Republican," a term which now seems quaint, in 1972, Mr. Lehmann again worked very successfully at a professional level in California politics; the California race was conducted with honor and success.

In late 1972, disenchanted with partisan politics, and keeping an oath made after the self inflicted death of a close friend resultant from depression after small weight felony arrests for marijuana, Mr. Lehmann joined former Reagan and Nixon senior campaign professional Gordon S. Brownell, in drug reform efforts which led to greatly reduced and misdemeanor penalties for the same conduct where arrests had led to the loss of his friend.

Having accomplished this narrow goal, Mr. Lehmann then abandoned politics. Mr. Lehmann has as of November of 2009 practiced law for more than 32 years, with a specialty focus since 1983 in litigation involving engineering and scientific proof, including matters involving accident reconstruction in aircraft and vehicle crashes, where he was a long time student of famed reconstruction expert H. Howard Hasbrook, a pioneer of that field. The firm also conducted a specialty practice in level failures and engineering cases involving failed foundations and similar structural engineering issues.

Mr. Lehmann estimates that he has handled or personally supervised more than 1600 civil actions, including, for example, serving as Class Counsel in the Chevron Aviation Gasoline class action, which resulted in new engines for roughly 1647 aircraft and the resultant avoidance of many crashes. Mr. Lehmann then served as co-

Class Counsel in the Mobil AV-1 class action, resulting in major engine repairs or overhauls for 850 aircraft; from these two cases, in company with aircraft crash litigation, Mr. Lehmann has represented more aircraft owners in U. S. litigation than any lawyer in the United States. Long fascinated by his legal work relating to the sciences and engineering, Mr. Lehmann has so far won five patents; those interested in engine physics may find recently granted U. S. Patent 7,581,526 of interest.

About Green Swan & L.D.A.D. Inc.

Lehmann Device & Design Inc., a small California company founded in 2002 as a patent and IP development company, is dedicated to inventive works for the reduction of carbon based pollution, and the pursuit of other environmental and public safety goals. The legal name change process is now underway in California for renaming LDAD Inc. to Green Swan Inc., to signal both the belief that well studied and careful civil dialogue can have an out-of-scale positive impact upon the human condition when focused with precision on the issues where the greatest benefit may be done, and that great focus is now appropriate around "Green" issues: *No Canary in the Quanta* is the first composition product of Lehmann Device & Design Inc., in said Green Swan spirit.

BIBLIOGRAPHY

"Gauss, Carl Friedrich." *Scientists: Their Lives and Works*, Vol. 107, Online Edition, 2006. Reproduced by Biography Resource Center, Farmington Hills, Mich. : Gale, 2008.

"LaPlace, Pierre-Simon." *Notable Mathematicians*, Gale Research, 1998. Reproduced in Biography Resource Center, Farmington Hills, Mich. : Gale, 2008.

"Scientists find 'local' back hole that pumps energy when it spins." Massachusetts Institute of Technology News Office, May 1, 2002. Online at: http://web.mit.edu/newsoffice/2002/blackhole-0501.html

"The Future Divined by the Crowd." *New York Times*, March 11, 2006.

"Upset that cosy Internet world." *New Scientist*. August 1, 2009.

Bernstein, Peter. Against *the Gods, The Remarkable Story of Risk*. New York : John Wiley & Sons, 1996.

Bohm, David. *Wholeness and the Implicate Order*. New York : Routledge Classics, 2002.

Browne, Malcolm. "A Bet on a Cosmic Scale, and a Concession, Sort of." *The New York Times*, February 12, 1977.

Cartright, Jon. "Fusion in a cold climate." *New Scientist*. July 18, 2009 (interview with Martin Fleischmann).

CERN official website, online at: http://public.web.cern.ch

Chomsky, Noam. *Chronicles of Dissent*. Monroe, ME : Common Courage Press, 1992.

Close, Frank. "Fears over Factoids". *Physics World*, August 2007. Online at: http://www.physicsworld.com

Dar, Arnon, A. De Rujula, Ulrich Heinz. "Will relativistic heavy-ion colliders destroy our planet?" *Physics Letters B* 470 (1999), 142-148.

Dash, Mike. *Tulipomania – The Story of the World's Most Coveted Flowers*. New York : Three Rivers Press, 2000.

Dylan, Bob. "To Ramona", *Another Side of Bob Dylan*, 1964.

Einstein, A., B. Podolosky and N. Rosen. "Can quantum-mechanical description of physical reality be considered complete?", *Physics Review*, 47 (1935).

Einstein, Albert. *Relativity: The Special and General Theory.*

Translated by Robert W. Lawson. New York : Random House, 1961.

Ellul, Jacques. *Propanga: The Formation of Men's Attitudes*. New York : Random House, 1965.

Everett, H. "Relative State Formulation of Quantum Mechanics." *Review of Modern Physics* 29, 1957.

Frankel, Mark. "When The Bubble Burst" (book review). *Business Week*, April 2000.

Gladwell, Malcolm. *Outliers*. New York : Little Brown and Company, 2008.

Gleick, James. *Chaos: The Making of a New Science*. New York : Penguin Non-Classics, 1988.

Goldgar, Anne. *Tulipmania: Money, Honor, and Knowledge in the Dutch Golden Age*. Chicago: University of Chicago Press, 2008.

Grann, David. "Trial by Fire." *The New Yorker*, September 7, 2009.

Greene, Brian (interview) at:
http://blogs.wnyc.org/radiolab/2009/03/25/diy-universe

Greene, Brian. *The Fabric of the Cosmos: Space, Time, and the Texture of Reality*. New York : Random House, 2004.

Haisch, Bernard. *The God Theory*. San Francisco, CA : Weiser Books, 2009.

Hoffer, Eric. *The True Believer: Thoughts on the Nature of Mass Movements*. New York : HarperCollins, 2002.

Huff, Darrell. *How to Lie with Statistics*. New York : W. W. Norton & Company, 1993.

Isaacson, Walter. *Einstein*. New York : Simon & Schuster, 2007.

Jaffe, R. L., W. Busza, and F. Wilczek. "Review of speculative 'disaster scenarios' at RHIC." *Review of Modern Physics*, Vol. 72, No. 4, October 2000, 1125-1140.

Johnson, Paul. Modern *Times Revised Edition: The World from the Twenties to the Nineties*. New York : Harper Perennial Modern Classics, 2001.

Kaku, Michio. *Parallel Worlds: A Journal through Creation, Higher Dimensions, and the Future of the Cosmos*. New York : Random House, 2005.

Kramer, Peter D. *Listening to Prozac*. New York : Viking Adult, 1993.

Lehmann, Harry V. "No Canary for the Quanta" (first essay), 2008.

Lehmann, Harry V. Letter to Modern Review of Physics after

submission of first No Canary essay.

Little, Lewis E. *The Theory of Elementary Waves*. Gainsville, Georgia: New Classics Library, 2009.

London, Jack. *The Call of the Wild*. Chicago : Nelson-Hall, 1980.

Lorenz, Edward N. "Deterministic Nonperiodic flow." *Journal of the Atmospheric Sciences* 20, No. 2, 1963.

Lorenz, Edward N. *The Essence of Chaos*. Seattle, WA : University of Washington Press, 1993.

Mackay, Charles. *Extraordinary Popular Delusions and the Madness of Crowds*. New York, Wiley, 1995.

Mallett, Ronald L. *Time Traveler: A Scientist's Personal Mission to Make Time Travel a Reality*. New York : Thunder's Mouth Press, 2006.

Mayer-Schonberger. "Can We Reinvent the Internet?" *Science*. Vol. 325, No. 5939. July 24, 2009.

McTaggart, Lynne. *The Field: The Quest for the Secret Force of the Universe*. New York : HarperCollins, 2003.

Miller, Dayton. "The Ether-Drift Experiments in 1929 and Other Evidences of Solar Motions." *Review of Modern Physics*, vol. 5(2), July 1933.

Nadeau, Robert and Menas Kafatos. *The Non-Local Universe: The New Physics and Matters of the Mind*. New York : Oxford University Press, 2001.

Newberg, Andrew, and Eugene d'Aquili. *Why God Won't Go Away*. New York : Ballantine Books, 2002.

Overbye, Dennis. "Asking a Judge to Save the World, and Maybe a Whole Lot More." *The New York Times*, March 29, 2008.

Pacella, Rena Marie. "Gambling On a God Particle." *Popular Science*, January 2008.

Packard, Vance. *The Hidden Persuaders*. Brooklyn, NY : Ig Publishing, 2007.

Penrose, Roger. *The Emperor's New Mind*. Oxford, England : Oxford University Press, 1999.

Radin, Dean. *Entangled Minds: Extrasensory Experiences in a Quantum Reality*. New York : Simon & Schuster, Inc.

Radin, Dean. *The Conscious Universe*. New York : HarperCollins, 1997.

Randall, Lisa. *Warped Passages: Unraveling the Mysteries of the Universe's Hidden Dimensions*. New York : HarperCollins,

2005.

Rees, Martin. *Our Final Hour.* New York : Basic Books, 2003.

Rosenblum, Bruce and Fred Kuttner. *Quantum Enigma.* New York : Oxford University Press, 2006.

Sardar, Ziauddin and Iwona Abrams. *Introducing Chaos.* Royston, UK : Icon Books, 2005.

Sheldrake, Rupert. *The Presence of the Past.* Rochester, VT : Park Street Press, 1995.

Skinner, B. F. *Beyond Freedom and Dignity.* Indianapolis, IN : Hackett Publishing Co., 2002.

Smolin, Lee. *Three Roads to Quantum Gravity.* Basic Books, 2001.

Sowell, Thomas. *The Vision of the Anointed.* New York : Basic Books, 1995.

Talbot, Michael. *The Holographic Universe.* New York : HarperCollins, 1992.

Taleb, Nassim N. *Fooled by Randomness: The Hidden Role of Chance in Life and in the Markets.* New York : Random House, 2005.

Taleb, Nassim N. *The Black Swan: The Impact of the Highly Improbable.* New York : random House, 2007.

Targ, Russell. *Limitless Mind, a Guide to Remote Viewing and Transformation of Consciousness.* Novato, California: New World Library, 2004.

Wiener, Norbert. *The Human Use of Human Beings: Cybernetics and Society.* New York : Houghton Mifflin, 1988.

Witt, Terence. *Our Undiscovered Universe: Introducing Null Physics.* Melbourne: Aridian Publishing Corporation, 2007.

Woit, Peter. *Not Even Wrong.* New York : Basic Books, 2006.

Wolf, Fred Alan. *Mind into Matter: A New Alchemy of Science and Spirit.* Portsmouth, NH : Moment Point Press, 2001.

Wouk, Herman. *A Hole in Texas: A Novel.* Boston : Little, Brown & Company, 2004.

END NOTES

Note: In some instances, where a referenced work cited by quotation contains one or more footnotes, those footnotes within the quotation may be omitted. Where this occurs, the abbreviation "(fn)" is supplied, and the reader may go to the cited original work for such detail.

[1] Rees, Martin. *Our Final Hour.* New York : Basic Books, 2003, 124. [Known in the UK as *Our Final Century]*

[2] Rees, Martin, 2003 (from Prologue).

[3] Jaffe, R. L., W. Busza, and F. Wilczek. "Review of speculative 'disaster scenarios' at RHIC." *Review of Modern Physics*, Vol. 72, No. 4, October 2000, 1125-1140.

[4] Jaffe, R. L., 2000.

[5] Lehmann, Harry V. "No Canary in the Quanta" (First circulated essay May 16, 2008).

[6] Lehmann, Harry V., 2008.

[7] Johnson, Paul. *Modern Times Revised Edition: The World from the Twenties to the Nineties.* New York : Harper Perennial Modern Classics, 2001.

[8] *The Borg* is a society of robotic fictional characters who think strictly as a 'group' and where there are no individuals; created within the television program "Star Trek Next Generation".

[9] Mayer-Schönberger, Viktor. "Can We Reinvent the Internet?". *Science.* Vol. 325, No. 5939. July 24, 2009, 396-397.

[10] "Upset that cosy Internet world." *New Scientist.* August 1, 2009, 17.

[11] Gladwell, Malcolm. *Outliers.* New York : Little, Brown and Company, 2008, 47-48.

[12] Gladwell, 2008, 49.

[13] Isaacson, Walter. *Einstein.* New York : Simon & Schuster, Inc., 2007.

[14] Kramer, Peter D. *Listening to Prozac.* New York : Viking Adult, 1993.

[15] London, Jack. *The Call of the Wild.* Chicago : Nelson-Hall, 1980.

[16] Chomsky, Noam. *Chronicles of Dissent.* Monroe, ME : Common Courage Press, 1992.

[17] Sowell, Thomas. *The Vision of the Anointed.* New York : Basic Books, 1995.

[18] Sowell, 1995.

[19] Sowell, 1995, 10.

[20] Sowell, 1995, 13.

[21] Taleb, N. *The Black Swan: The Impact of the Highly Improbable.* New York : Random House, 2007.

[22] Taleb, N., 2007.

[23] Dash, Mike. TULIPOMANIA - The Story of the World's Most Coveted Flower. New York : Three Rivers Press, 2000.

[24] Dash, 2000.

[25] Dash, 2000.

[26] Newberg, Andrew, and Eugene d'Aquili. *Why God Won't Go Away.*

[27] "Scientists find 'local' black hole that pumps energy when it spins." Massachusetts Institute of Technology News Office, May 1, 2002. Online at: http://web.mit.edu/newsoffice/2002/blackhole-0501.html

[28] Overbye, Dennis. "Asking a Judge to Save the World, and Maybe a Whole Lot More." *The New York Times*, March 29, 2008.

[29] Overbye, 2008.

[30] MIT News Office, 2002.

[31] MIT News Office, 2002.

[32] Radio interview with Brian Greene (podcast), online at: http://blogs.wnyc.org/radiolab/2009/03/25/diy-universe

[33] Pacella, Rena Marie. "Gambling On a God Particle." *Popular Science*, January 2008, 46.

[34] Taleb, Nassim N. *Fooled by Randomness: The Hidden Role of Chance in Life and in the Markets.* New York : Random House, 2005, 105.

[35] Wiener, Norbert. *The Human Use of Human Beings: Cybernetics and Society. New York : Houghton Mifflin*, 1988, 7.

[36] Bernstein, Peter Bernstein. *Against The Gods, The Remarkable Story of Risk.* New York : John Wiley & Sons, 1996.

[37] "LaPlace, Pierre-Simon." *Notable Mathematicians*, Gale Research, 1998. Reproduced in Biography Resource Center, Farmington Hills, Mich. : Gale, 2008.

[38] "Gauss, Carl Friedrich." *Scientists: Their Lives and Works,* Vol. 107, Online Edition, 2006. Reproduced in Biography Resource Center, Farmington Hills, Mich. : Gale, 2008

[39] Wiener, 1988, 7-10.

[40] Wiener, 1988, 10.

[41] Smolin, Lee. *Three Roads to Quantum Gravity.* Basic Books, 2001, xxi.

[42] Lorenz, Edward N. The Essence of Chaos. Seattle, WA :

University of Washington Press, 1993.

[43] Mallett, Ronald L. *Time Traveler: A Scientist's Personal Mission to Make Time Travel a Reality.* New York : Thunder's Mouth Press, 2006, 21 quoting from Bradbury's "A Sound of Thunder", *Colliers*, June 28, 1952.

[44] Lorenz, Edward N. "Deterministic Nonperiodic Flow." *Journal of the Atmospheric Sciences* 20, no. 2 (1963); 130-141.

[45] Dylan, Bob. "To Ramona", *Another Side of Bob Dylan*, 1964.

[46] Taleb, 2007, 138.

[47] Taleb, 2007, 151.

[48] Taleb, 2007, 154.

[49] Taleb, 2007, 161.

[50] Taleb, 2007, 161.

[51] Taleb, 2007, 243.

[52] Gleick, 1987, 125-127, 130-138.

[53] Taleb, 2007, 161.

[54] Taleb, 2007, 243.

[55] Dar, Arnon, A. De Rújula, Ulrich Heinz. "Will relativistic heavy-ion colliders destroy our planet?" *Physics Letters B* 470 (1999), 142-148.

[56] Everett, H. "Relative State Formulation of Quantum Mechanics." *Review of Modern Physics* 29 (1957): 454-462.

[57] Kaku, Michio. *Parallel Worlds: A Journal through Creation, Higher Dimensions, and the Future of the Cosmos.* New York : Random House, 2005.

[58] Penrose, Roger. *The Emperor's New Mind.* Oxford, England: Oxford University Press, 1999, 362-363.

[59] Radin, Dean. *The Conscious Universe*, New York : HarperCollins, 1997, 287.

[60] Smolin, 2001, 33-34.

[61] Greene, Brian. *The Elegant Universe: Superstrings, Hidden Dimensions, and the Quest for the Ultimate Theory.* New York: W.W. Norton & Company, 1999, 3.

[62] Greene, Brian. *The Fabric of the Cosmos: Space, Time, and the Texture of Reality.* New York : Random House, 2004, 376.

[63] Lorenz, 1963.

[64] Witt, Terence. *Our Undiscovered Universe: Introducing Null Physics.* Melbourne: Aridian Publishing Corporation, 2007.

[65] Witt, Terence, 2007, 13.

[66] Smolin, 2001, 209.

[67] Randall, Lisa. *Warped Passages: Unraveling the Mysteries of the*

Universe's Hidden Dimensions. New York : Harper Collins, 2005, 283.

[68] Randall, 2005, 211.

[69] Woit, 2006.

[70] Woit, 2006 (Krauss quote found on book's jacket).

[71] Woit, 2006, 210.

[72] Hoffer, Eric. *The True Believer: Thoughts on the Nature of Mass Movements.* New York : HarperCollins, 2002, 63.

[73] Randall, 2005, 67.

[74] Johnson, 2001, 718.

[75] Woit, 2006, 207.

[76] Radin, Dean. *The Conscious Universe*, New York : HarperCollins, 1997, 56.

[77] Radin, 1997, 57.

[78] Radin, 1997, 58.

[79] Radin, 1997, 57-58.

[80] Johnson, *Modern Times*, 2001, 2.

[81] Einstein, Albert. *Relativity: The Special and General Theory.* Translated by Robert W. Lawson. New York : Random House, 1961, 147.

[82] Miller, Dayton. "The Ether-Drift Experiments in 1929 and Other Evidences of Solar Motion." *Reviews of Modern Physics,* Vol. 5(2); July 1933.

[83] Einstein, 1935.

[84] Radin, Dean. *Entangled Minds: Extrasensory Experiences in a Quantum Reality.* New York : Simon & Schuster, Inc., 2006, 1.

[85] Radin, 2006, 227.

[86] McTaggart, Lynne. *The Field: The Quest for the Secret Force of the Universe.* New York : HarperCollins, 2003.

[87] Radin, 2006, 95-97.

[88] Targ, Russell. *Limitless Mind, a Guide to Remote Viewing and Transformation of Consciousness.* Novato, California : New World Library, 2004, 25-27.

[89] Bohm, David. *Wholeness and the Implicate Order.* New York : Routledge Classics, 2002.

[90] Randall, 2005, 445.

[91] Wolf, Fred Alan. *Mind into Matter: A New Alchemy of Science and Spirit.* Portsmouth, NH : Moment Point Press, 2001, 75.

[92] Radin, 1977, 287.

[93] Little, Lewis E. *The Theory of Elementary Waves.* Gainsville,

Georgia: New Classics Library, 2009, 7.

[94] Isaacson, 2007, 164.

[95] A reader's review of Dr. Little's book, *The Theory of Elementary Waves*, as posted on Amazon, September 2009.

[96] Randall, 2005, 293.

[97] Kaku, 2004, 353.

[98] Mallett, 2006, xii-xiii.

[99] Lehmann, Harry. Letter to Modern Review of Physics, 2008.

[100] Kaku, 2005.

[101] Schneider, Frank. Quote from dust jacket of Lewis E. Little's *The Theory of Elementary Waves*.

[102] McKerrell, Archie. Quote from dust jacket of Lewis E. Little's *The Theory of Elementary Waves*.

[103] Flagg, Michael. Quote from dust jacket of Lewis E. Little's *The Theory of Elementary Waves*.

[104] Little, 2009, xiii.

[105] Little, 2009, xvii.

[106] Little, 2009, xiii.

[107] Little, 2009.

[108] Little, 2009.

[109] Little, 2009, 36.

[110] Sheldrake, Rupert. *The Presence of the Past*. Rochester, VT : Park Street Press, 1995.

[111] Sheldrake, 1995, 300.

[112] Little, 2009, 18.

[113] Radin, 2006.

[114] Grann, David. "Trial by Fire." *The New Yorker*. September 7, 2009.

[115] Judicial Council of California, Section 223. Published online by LexisNexis Mathew Bender, 2009.

[116] Judicial Council of California, Section 224. Published online by LexisNexis Mathew Bender, 2009.

[117] Grann, 2009.

[118] Grann, 2009.

[119] Grann, 2009.

[120] Gladwell, Malcolm. *Outliers*. New York : Little, Brown & Company, 2008.

[121] "The Future Divined by the Crowd". *New York Times*, March 11, 2006.